LEISURE / TOURISM GEOGRAPHIES

Leisure / Tourism Geographies considers leisure/tourism as an encounter – an encounter that exists between people, people and space, people as socialised and embodied subjects, expectations, experience and desire.

The contributors explore diverse aspects of leisure and tourism, ranging from the methodologies behind leisure practices to detailed case studies, which include: Disneyland Paris; tourism in sacred landscapes; leisure practices in cyberspace; leisure and yachting; consuming pleasures; the use of recreational/holiday cottages; National Parks; and local parks and gardens.

Presenting an exciting mix of attitudes and ideas concerning leisure and tourism, this book documents a lively debate, placing geography at its centre.

David Crouch is Professor of Cultural Geography at Anglia University and visiting Professor of Geography and Tourism at Karlstad University, Sweden.

CRITICAL GEOGRAPHIES
Edited by Tracey Skelton
Lecturer in International Studies, Nottingham Trent University
and
Gill Valentine
Senior Lecturer in Geography, The University of Sheffield.

This series offers cutting-edge research organised into three themes: concepts, scale and transformation. It is aimed at upper-level undergraduates and research students, and will facilitate inter-disciplinary engagement between geography and other social sciences. It provides a forum for the innovative and vibrant debates which span the broad spectrum of this discipline.

LEISURE/TOURISM GEOGRAPHIES

Practices and geographical knowledge

Edited by David Crouch

London and New York

First published 1999
by Routledge
11 New Fetter Lane, London EC4P 4EE

Simultaneously published in the USA and Canada
by Routledge
29 West 35th Street, New York, NY 10001

Routledge is an imprint of the Taylor & Francis Group

Typeset in Perpetua by
The Florence Group, Stoodleigh, Devon.
Printed and bound in Great Britain by
St Edmundsbury Press, Bury St Edmunds, Suffolk.

British Library Cataloguing in Publication Data
A catalogue record for this book is available
from the British Library

Library of Congress Cataloging in Publication Data
A catalogue record for this book has been applied for.

ISBN 0–415–18108–9 (hbk)
ISBN 0–415–18109–7 (pbk)

CONTENTS

CONTENTS

PLATES

CONTRIBUTORS

Cara Aitchison is Senior Research Fellow in the Leisure and Sport Research Unit at Cheltenham and Gloucester College of Higher Education, UK. She is co-author of *Leisure and Tourism Landscapes: social constructions of space and place* (Routledge 2000).

John Bale is Professor of Sport Geography at Keele University, UK.

Caroline Bassett is Lecturer in Media Studies at Sussex University, UK.

Inger Birkeland is Research Fellow in Geography at the University of Oslo, Norway.

Mike Crang lectures at the University of Durham and is author of *Cultural Geography* (Routledge 1998).

David Crouch is Professor of Cultural Geography at Anglia University, UK and Visiting Professor of Geography and Tourism, Karlstad University, Sweden. He is co-editor of *Cultural Turns* (Longmans 2000), and author, with Colin Ward, of *The Allotment: its landscape and culture* (Faber and Faber 1988).

Gert Groening is Professor of Landscape Architecture at the Institute for the History and Theory of Design, Hochschule der Kuenste, Berlin, Germany.

Lena Jarlöv is a Research Fellow at the Dalarnas Forskningsrad Research Institute, Falun, Sweden.

Bjørn P. Kaltenborn is Senior Research Scientist in Tourism and Environmental Planning at Eastern Norway Research Foundation, Lillehammer, Norway.

Niels Kayser Nielsen lectures in Cultural History at the University of Odense, Denmark.

Ada Kwiatkowska lectures in Architecture at the Wroclaw University of Technology, Poland.

Eric Laurier is Research Fellow in Geography at Glasgow University, Scotland.

Deborah Philips lectures in English at Brunel University, UK. She is co-author of *New Causes* (Cassell 1999) and *Writing Well* (Kingsley 1999).

Neil Ravenscroft is Reader in Leisure Management at the University of Surrey, UK.

Uwe Schneider lectures in Landscape History at the Hochschule der Kuenste, Berlin, Germany.

Richard Tresidder is completing a PhD in tourism and representations and the practice of Cornish identities in the Tourism Department at Derby University's Business School, UK.

John Urry is Professor of Sociology at Lancaster University, UK, and author of *The Tourist Gaze* (Sage 1990), *Consuming Places* (Routledge 1995), co-author with Phil MacNaghten of *Contested Natures* (Sage 1998), and co-editor with Chris Rojek of *Touring Cultures* (Routledge 1997).

Gill Valentine lectures in Geography at Sheffield University, UK and is co-author with David Bell of *Consuming Geographies* (Routledge 1997) and co-editor with Tracey Skelton of *Cool Places* (Routledge 1998), and with David Bell of *Mapping Desire* (Routledge 1995).

Stacy Warren lectures in Geography at Eastern Washington University, USA.

Chris Wilbert is Research Fellow at the City University, London (UK) Department of Social and Human Sciences. He is co-editing a book on animal geographies with Chris Philo.

Daniel R. Williams is currently Research Social Scientist at the Rocky Mountain Research Station of the United States Forest Service, Fort Collius, Colorado, USA. His contribution to this volume was written while he was Associate Professor of Leisure Studies, University of Illinois, USA.

ACKNOWLEDGEMENTS

The authors individually make their own acknowledgements. Many colleagues have contributed to this developing project that includes discussions at many universities and conferences in Europe and the USA, and at the Leisure Geographies Workshop, jointly shared by the Institute of British Geographers social and cultural geography study research group (IBG/RGS) and the Leisure Studies Association. I thank all of these and all of the authors for their working practice, and Routledge editors for their positive feeling about the project from its inception.

David Crouch

1

INTRODUCTION: ENCOUNTERS IN LEISURE/TOURISM

David Crouch

Introduction

This book considers leisure/tourism as an encounter. That encounter occurs between several things. It occurs between people, between people and space, amongst people as socialised and embodied subjects, and in contexts in which leisure/tourism is available. The encounter is also between expectations and experience, desire, and so on. This chapter introduces these themes. In their own way each chapter explores different ways in which people make leisure/tourism. Thus aspects of the encounter are actively contextualised and practised by active human subjects. People as human beings (rather than, say, consumers) enjoying leisure/tourism are the subject of this book. Leisure/tourism is a process rather than a product, although it is frequently abstracted as the latter. Knowledge in leisure/tourism is also a matter of process, being worked, refigured, in flows rather than fixed in time-space. As a practice leisure/tourism involves complex human and social engagements, relations and negotiations and it is therefore appropriate to write in terms of 'practice'. Of course, leisure/tourism may not become enjoyment and instead may be marked by a frustrated hope or mirage and negative memory. Space is important in each of its aspects.

Tourism is often theorised in terms of meetings of cultures, or one-way exploration, and suggests travel. There is a persistent confusion of categories between leisure and tourism. This is no accident, as tourism and leisure become de-differentiated in post-Fordism and together are emblematic of postmodernity (Lash and Urry 1994). As tourism *and* leisure have become less and less functional and increasingly aestheticised, the differentiation of tourism and leisure is eroded. Leisure becomes commodified (in places) and tourism is accompanied by similar commodification, and both have capacity for reflexivity. This book adopts an ambivalent position and uses the couplet 'leisure/tourism'. Distinctive chapters take up aspects that may be recognised as more one than the other. In terms of practice there are more similarities than differences. This book deals with both – or the one, sometimes dual, category.

Practices that we understand by leisure/tourism merge with other areas of life, and work regimes are becoming increasingly flexible. For many people their primary investment of themselves and their identity may still be in their work. It may be that what they do outside of work is merely routine (house-keeping, garden maintenance, 'necessary' visits to family, looking after the children) yet for others may be key fulfilling parts to their lives as leisure. Leisure may become similarly imprecise (entertaining and playing golf with colleagues for example). Supporting services of leisure/tourism can double for work, such as filling stations, airport lounges (Augé 1995). Strictly, time outside unpaid work as a distinction of leisure is equally misleading, as numerous feminist texts have shown. Leisure/tourism is a means of practising space, although not the only one, as Crang observes in terms of work (1994). Leisure/tourism can provide one means of imaginative play and expressive behaviour and identity.

The individual is relatively grounded in different sets of context: as socialised subject, exemplified in this volume as the gendered subject; the socialised contexts of particular leisure spaces and practices and abstract ideas of space, practice, and leisure/tourism. Just as leisure/tourism is very contextualised, socially and culturally, we too are profoundly socialised subjects. In Bourdieu's terms, our *habitus* of settings and contexts are sustained by practice (1984, 1990). In the contemporary period however, the possibility of being reflexive subjects unsettles this relation (Crouch and Matless 1996). Indeed, Lash and Urry have argued that leisure/tourism is a typical example of the reflexivity of contemporary culture, with opportunities for aestheticisation, in a critical negotiation of the self (1994).

Space and knowledge

Leisure/tourism happens in spaces. That space may be material, concrete and surround our own bodies. Space may be metaphorical and even imaginative. That imaginative space is not only in contemporary virtual leisure but in the imaginative practice of the subject. Metaphorical space in leisure includes abstractions of city and country, nature, and many leisure/tourism practices happen in spaces that are culturally defined. In the more poetic sense of Bachelard, abstract notions and senses of sites of being – corners, enclosures, 'home' – may be inflected by an actual knowing of spaces in material, everyday encounters (1994). Thus space is important in metaphorically 'shaping' the enjoyment of leisure. Such an interpretation of the importance of space in leisure/tourism differs markedly from a more instrumental, empiricist view of space as enabling leisure/tourism to be 'located'. These chapters consider the encounter of leisure/tourism as practised.

In the process of enjoying leisure/tourism as human subjects we figure and refigure our knowledge of material and metaphorical spaces where that leisure happens. Particular features of geography become artefacts through which we remember. They are also embodied in our knowledge as lay geographies that enfold and are enfolded by relationships, movement and laughter (Radley 1990).

There is a renewed interest in popular geographical knowledge as lay geographies. This approach to popular geographies works from the subject and practice rather than from representational geographies of media popular culture. However, this does not mean ignoring context and socialisation of the subject. Instead it seeks to wrestle with the active process of practice, where subjects may negotiate and refigure. Lay geographical knowledge amounts to an ontological knowledge, 'the feeling of doing' (Shotter 1993). Once again, this is not to imply that 'doing-knowledge' is wholly different from learnt, expert knowledge or of knowledge by reading representations. It is instead a knowledge that incorporates all of these, and is perhaps incoherent, certainly 'incomplete' and uneven. Harre's notion of the feeling of doing is not the same. Rather it emphasises those components of practice that directly, freely engage the body in the surrounding physicality of the world (Harré 1996). This acknowledges a sensual, multi-sensual subject (Pile and Thrift 1995) that comes to know the world in part at least as embodied subject.

Knowledge may be too strong a word to apply to the often fleeting, sensual practice of leisure/tourism. Movements and engagements may be vicarious. Bauman argues that play is not cumulative (Bauman 1993). Up to a point. Like the occupation of liminal spaces these moments of practice offer a refiguring of normal roles that may be played with or played at rather than temporarily erased or inverted (Urry 1990). Repetitive and ritualistic practice such as may accompany visits to the theme park and sports ground, as much as music group performance, offers time for mutual refiguring, reappraisal, or avoidance, of everyday knowledge. The complexity of these events, the intersubjective nature of leisure/tourism, the tracking of different cultural and social meanings and representations and the activity of the socialised, embodied subject coping, trying out, negotiating, contesting, makes the term practice very appropriate.

Part of the interest in reflexivity and in the non-representational has been the emergence of new interest in the poetic. Rojek argues that the poetic has been ignored for too long, yet has been used with an everyday acceptance in the literature that promotes tourism (1995). Travel is frequently aligned to the exotic, rendered in the poetic. Crang's case for lay geographies includes an explicit acknowledgement of the poetic (1996). Ann Game's discussion of de Certeau's working of spatial practice is important. Whilst de Certeau's words that 'every story is a travel story' would seem to support the organised romance of tourism companies, he emphasises that such a travel story is a *spatial practice* (de Certeau

1984:115, Game 1991:166). To make a spatial practice is to engage in a transformation, not to return or imagine a past, but creatively to enliven, to repeat
only the possibility of a new, unique moment. Agencies that represent tourism
and leisure can only provide structures into which our imaginative practice enters
and through which it explores its desires, and their promotional messages that
inflect these structures may not be ours. Crude consumption figures do not reveal
very much of spatial practice. Indeed, those structures may deflect or deaden
interest. We may need to come to a place unawares.

For Ann Game her 'desire has indeed been to know the place, to be able to
read the codes of, for example, public footpaths and bridle ways; to have a
competence with respect to this landscape as [she does] in body surfing at Bondi;
to be a local and party to local stories. In a sense, it is a desire to "know" what
cannot be seen' (1991:184). It may be easy to misrepresent what Game is saying.
She does not want to read according to a proscribed story, but from the plurality
of phenomena and imagination that may be engaged by letting oneself open to
space itself. This does not require one to walk, but to experience the surrounding
world. It is not a gaze but a multi-sensual sensitivity that is also in touch with
an imaginary sensitivity. Game does not romanticise de Certeau (who can overstate the socialised subject's freedom to do and to act and to transform), but
uses his text to raise the possibility of the subject from the familiar repressive
position of object. These dimensions of spatial practice are explored through the
case of visiting the Arctic Circle by Inger Birkeland (Chapter 2). Birkeland
explores the poetic and the working of myth in the subject. Engaging the gendered
subject and using Kristeva (1996) she explores and develops an interpretive study
of the feminine dimension of the human subject and the possibilities of imagination and myth in refiguring place.

The embodied subject is a very important part of Birkeland's work. Instead of
a seeing and objectifying practice, leisure/tourism emerges as a more feminine,
subjective, human spatial practice. One aspect of this is the increasing recognition
of the plurality of senses that give access to the world. In Chapter 3 John Urry
takes up the story on the gaze with an examination of smell as an informative
component of geographical knowledge. In practising space the imaginative,
sensual subject encounters metaphors as well as the material world on which
different discourses may be inscribed.

The country and the city, the garden, the beach, the desert island, and the
street hold powerful metaphorical attention in significant arenas of leisure/
tourism (Crouch 1998, Fyfe 1998, Crouch 1992, Selwyn 1996). Macnaghten
and Urry have demonstrated this with regard to the predominantly twentieth-
century and modern labelling and imagining of the countryside in terms of 'peace
and quiet', where the body rests (1998). Bale's Chapter 4 explores the use of
metaphor in the contexts of practice, through the example of sports stadia.

Popular thinking about leisure/tourism involves unexpected cross-overs and here we observe the overlaps of leisure site production between horticulture, stadium and cultivation. These properties and the use of metaphor are interpreted as highly gendered.

Patterns of investment, planning and promotion hold powerful influences over the context in which places, sites and cities are 'consumed'. It may be surprising that the modern assertion of masculinity in terms of stories of city 'fathers' told in the urban fabric is being recycled at the turn of the century. This gendering of places continues to contextualise leisure/tourism and to provide a theatre in which people may negotiate their own knowledge. In the absence of alternatives it is likely to be that city narratives like these will prolong the image of the heroic male. Aitchison's Chapter 5 considers the example of a city that combines a hegemonic story of heritage, gender and nation.

The modern way in which the state provided context for leisure/tourism was through controls and regulations that combined morality, class and gendering, influenced by socialised concern for health and environment, often in the exercise of power (Clarke and Critcher 1985, Rojek 1993, Revill and Watkins 1996). Popular political contest from people who, through leisure/tourism challenged these contexts, led to adjustments in control, exemplified in the Mass Trespass in the Peak District between the wars (Shoard 1997). Clubs have had their own influence on shaping the context of leisure/tourism amongst their members (Matless 1995). Interestingly, the Peak District became the site of the first 'National' Park a decade and a half later. More recently 'countryside' has been commodified (Crouch 1992) rather than democratised and this is demonstrated in a peculiar case of access for rambling. Ravenscroft's Chapter 6 explores how the management and control of freedom deploys modernist metaphors of countryside (purity, landed ownership) and its politics. However, in an effort to respond to demands for increased access it uses the virtual language of hyper-reality instead to *imply* access. Payments are apparently made to owners of land in a claim to provide access whose materiality is neither on the ground, nor on maps. The contexts for the subject, the *status quo* of control and regulation, remain unchanged and there is a reassertion of traditional countryside leisure access formed around traditional land ownership. This is accompanied by a new ideology of commodification as the appropriate means to achieve access.

The socialised subject copes or wrestles with or celebrates leisure/tourism sites culturally defined and embedded in metaphor. The chapter by Philips (Chapter 7) explores the resonance between the heavily, culturally signposted form of the theme park with the children's story from which it takes many of its cues. The space of the theme park is crowded with culturally significant representations. Sites like these become echoes of early socialisation rather than reproduce their own relations. Moreover, contents like these do not saturate the visitors' expectations and practice. These sites can provide a loose shell in which the subject makes her own meaning and knowledge through her own shared practice and imagination, exemplified in Ley and Olds' findings in a World's Fair, as a good day out for the family rather than an experience of global reference (1988).

Warren argued the importance of commodified leisure landscapes in contextualising practice, but also its limits (1993). People may consume, but they make their own sense and value, their own knowledge, albeit negotiated with a myriad of influences.

However, this issue of 'limits' becomes focused when those contexts are represented and performed in ways that are directly at odds with ways people want to practice space and the meanings and relationships they intend. In Chapter 8 in this volume Warren considers evidence of reaction and popular critique of Disneyland Paris that articulates negotiation with aspects of colonialism and gender.

Consumption is a useful theoretical tool, but may be inadequate to account for leisure/tourism's spatial practice. Finnegan rejects the 'productionist' view of consumption as shaped by production because it renders people standardised and passive, with little role for human agency (1989, 1997). She argues that considering consumption in terms of a definitive link to patterns of inequality and social differentiation, with consumers ultimately reproducing their class positions (strongly argued in Bourdieu's work on *habitus* (1984)), is over-simplistic. In her own work on leisure music-making she found both uses of 'consumption' incomplete (1989). Distinctions of class break down because people doing leisure frequently come from different social positions and backgrounds. Moreover, they may *produce* the cultural forms they enjoy. They may take fragments from produced goods and forms, but reassemble these significantly as they choose, where the familiar producer-consumer relation is unimportant. Consumption may be especially useful in understanding aspects of shopping (Shields 1992, Jackson and Thrift 1995, Miller *et al.* 1998). Increased recognition of active and more diverse modes of practice, at the limits of commercially related consumption, is helpful in loosening the hold of earlier consumption models (Gregson and Crewe 1997).

A walk in the country, a visit to a theme park or the local park or pub offer numerous spaces for people to create their own play, as well as enthusiasms in which people informally take part and self organise (Crouch and Tomlinson 1994). Finnegan found the notions of ritual, artistic convention and unique performance more useful in understanding what people do in other cases of leisure (1997). Although people may be influenced by holiday location brochures, the images may become merely imaginary diversions, indefinite mythical aspirations. Several chapters in this volume explore what happens when people do participate in commercialised leisure, in commodified spaces. A further distinction we may reconsider is that between performer and spectator.

The person 'spectating' at an historic enactment turns and engages, perhaps also imaginatively and emotionally, discusses events, meets with friends over tea. To reduce the spectator to a merely passive role omits an understanding of practice. As in the case of sports crowd control there is regulation in these spaces, but not one that is overpowering. Niels Kayser Nielsen has explored the role of the spectator in terms of spatial practice and embodiment. He argues that people move in surrounding space, may be aware of a 'buzz' of enthusiasm – or despair – in the crowd en route to or from the ground. In doing so they are constrained by regulation, and policing streets surrounding football grounds, for example, is now commonplace. However, he argues that even in this regulated space amongst relative 'strangers' there is a transformation (1995). Zukin finds that women in security-regulated city parks find greater confidence (1995). There is a crucial tension between contexts, socialisation and the subject's practice.

Practice offers a model of a more active process than consumption. This may apply in particular to expressive and performative leisure, in terms of 'ritual enactment' and 'art', rather than 'consumption' and 'power'. How far can such labels extend? Can we include 'just playing around', youth on a vacant site, at a theme park, walking, drinking together at the pub and sitting reading a Sunday newspaper in the car park (Crouch 1994, 1998)? To deny expressive faculty, imaginative power and a capacity for reflexivity in any of these implies elitism and thereby has little to do with understanding lay knowledge. These may offer important opportunities for self-empowerment and identity, creativity and confident self-feeling, which may also be shared with others, evident in the expression of 'ownership' by people of their local parks, for example (Crouch 1994), that has nothing to do with either legal or financial influence.

Pre-modern ritual was an important informing context of leisure/tourism in Poland. This was refigured, but not erased, by the imposition of glorified work and organised leisure of the Stalinist period. However, now the same sites are being refigured again, partly as a rediscovery of ritual, partly as a new attitude of life stylisation (Kwiatkowska Chapter 9). It is likely that people will renegotiate what they make of visits and journeys in their new commercial wrapping. It is unlikely that they will merely replace a valued ritual, already upheld amidst considerable adjustments over half a century. It is equally unlikely that for a new generation they will remain exactly the same, but this new generation has an understanding of these practices that they have heard from, and perhaps shared with their parents. An orientation towards practice enables more rigorous thinking about the complexities rather than the gross changes involved in remaking leisure/tourism. Tourism in contemporary Western culture struggles to cope with traditional transference too. Tourism spaces are still imagined and practised as sacred, but in new, sometimes ironic ways. Richard Tresidder explores this complexity (Chapter 10), and returns us to important questions of reflexivity and of how seriously we take tourism brochures.

Embodiment

In Western Europe gardening exemplifies the arena of negotiation and the representation of leisure/tourism as commodity life stylisation. It is being transformed into a series of commodities linked closely with lifestyle. There is a new wave of gardening programmes on UK television concerned with redesigning gardens promoted as 'style changes'. Interestingly, there has been another series called 'Paradise Gardens' (Catalyst 1996). This series invokes the metaphor or myth of paradise and reminds us that the garden can be envisaged, imagined, and even practised, as a liminal space. The possibilities of the poetic in gardening represent a real problem for those involved with its design as a moral space of order, as typified by formal designing in the early twentieth century, even with zoned consumption areas. In Chapter 11, Groening explores a Modernist construction of the garden that dominated much of twentieth-century garden design in contrast with its practice, and includes contemporary evidence. In adjusting or replanting a garden, however humbly, there is a negotiation with available influences and contexts such as those presented in popular magazines. The domestic gardener negotiates and wrestles with these contexts in an expressive practice of individuality. The garden is a site of embodied practice, where the body moves laterally, multi-dimensionally, bending, turning, engaging its body space in numerous, often tactile and multi-sensual ways. Performance offers a strong example of imaginative practice.

Expressive practice in leisure/tourism happens in very diverse ways. In dancing these components are perhaps more explicit. The body conveys expressivity and extends the self to relate to surrounding objects, whose meaning is transformed according to the feeling of movement and imagination in a particular space (Radley 1995, Thrift 1997). This individual and creative expressivity does not efface the socialised subject, but nor does the reverse occur (Radley 1996). The body practices and performs in a body-related space. This is likely at least partially to influence the way the subject makes sense of space, despite claims of the erasure of 'real' space in virtual 'worlds'.

The socialised subject can deploy expressive tactics in subverting, or merely in unsettling contexts, reflexively making them work. Both the sites and practices of consumption are given gendered and sexualised metaphors. This can construct uneven power relations between subjects, and it is crucial in making sense of lay spatial practices to interpret how the subjects together may handle these. Gill Valentine develops an argument on leisure practice through embodied negotiation as gendered, sexualised subjects (Chapter 12). The embodied subject does not only behave in ways contextualised in commodified representations. Practice confers rights over the body that may be very different from those contextually asserted. Indeed, the highly regulated practice of sports people, professional or not, is only partial (Bale 1993). Merleau-Ponty writes of body-space emphasising the importance of empowerment and Bourdieu balances this by reminding us that sport practice is highly regulated but the player still has room to adjust (Merleau-Ponty 1962, Bourdieu 1990). He does so not with an unlimited reading of controls over movement in sport, but by pointing to the tactics and inter-relation between rules and freedom, or contexts and reflexivity. De Certeau argues an ever-present possibility of liberation through the way we use space (1984). Young has argued a distinctive feminine practice of space through 'throwing like a girl'. However, she adds that whilst she maintains the distinctively socialised confinement and regulation of the female – as indeed male – body, strong reason to modify the application of the de Certeauian transforming walker, her discussion of throwing like a girl may understate a woman's subjectivity (1990:11). Whether female or male, of course gendered, there is a negotiating, a wrestling with that gendering and the often gendered contexts of spaces of practice. The borders and boundaries are blurred and indistinct, and it is to the challenge of investigating this bordering that this volume points direction.

Warren's chapter develops further the idea of reflexive negotiation and the possibility of refiguring in ways that make their own meaning of events and their own knowledge. Disneyland Paris is of course heavily contextualised but interacts with other knowledges (Chapter 7), and is frequently used to exemplify commodification, representation and the gaze. However, here we understand the subject to be engaged in working other knowledge to contest the play of representations and in particular their gendered and colonial contexts. Working a very different example, Chapter 13 by Bassett and Wilbert considers the power of commodification and of the subject, rather than the engagement by the subject in practice in order to counter exaggerated claims of virtual leisure's early freedoms.

The socialised subject is also embodied emotionally, and emotion is an important influence on the practice of learnt skill. Shotter argues that the subject makes a working of skill knowledge through practical, theoretical and in his particular terms, *participatory knowledge*. The latter is linked to social and personal identities (1993:7). This tension between emotion and skill is the subject of Laurier's Chapter 14 on elite leisure. Here elite knowledge is embodied, working learnt skill with emotion into a very distinctive kind of lay geography.

Embodiment is important in the way people seek to re-centre themselves through leisure/tourism and use space in the process. Whilst the postmodern de-centred subject has significantly refigured the way we practice leisure/tourism (Rojek 1995), there is a continual effort to re-centre in terms of identity (Giddens 1991), and space can be a very useful tool for transforming apparently distanciated selves into the embodied and familiar. This may be a nostalgic or imaginative process. It may involve dimensions of human subjectivities frequently omitted from leisure/tourism theory. These are posed by Gorz: 'Communication, giving, creating and aesthetic enjoyment, the production and reproduction of life, tenderness, the relation of physical, sensuous and intellectual capacities, the creation of non-commodity values' (1984).

Finding and reworking new social relationships and identities are important components of practice. Even the minutiae of events, enthusiasms and sites become important means of making social relations (Moorhouse 1992). Again, space is important in this process. Maffesoli argues that space is a convenient focus of socialities because people tend to get together in particular places, they walk to the pub, they may drive to a rehearsal for the evening but not too far (Maffesoli 1996, Finnegan 1989). There is, then, an influence of spatial proximity, 'proxeme', that associates space and practice. Events happen, indoors or outdoors, in the street, across the theme and city park, on the beach, in particular spaces and these spaces also figure in the lay geography of practice. In Radley's terms they configure our memory, as embodied in our practice. In refiguring social relations there is a frequent investment of self-generation in leisure/tourism, fairly detached from consumerist conditions (as in the case of the car-boot fair), and sometimes even outside the regulation of the state (Bishop and Hoggett 1986). More likely people assert themselves despite such regulations and contexts and socialised selves, unevenly and uneasily working their practices and knowledges, as several chapters show. These tactics may be regarded as typical means of reflexivity.

Whilst opportunities abound for local exclusionary feeling and inward looking identities and practices, essentialist in a Heideggerian way, these may also be progressive (Massey 1993). Nature can be refigured as an imaginative site in which other relations can be refigured too. These possibilities are taken up in a case study of summer cottaging by Williams (Chapter 15). Human, subjective creativity is an influence on the way cottage-holders make their encounters in practice. Creativity is a theme that is considered in Chapter 16 by Jarlöv. In her consideration of allotments there is a keen comparison with cottaging or spending time in a summer house. In each she argues there is a metaphorical and material space for subjectivity, empowerment, identity building. These practices are worked through a creativity that is also expressive. The examples considered in these two chapters address very different groups, one relatively middle class and the other significantly 'working class', yet through these examples we discover a process of subjectivity which works with particular contexts. Whilst differently socialised and limited by their material resources as well as by their disposition and *habitus*, what people do bears similar characteristics of negotiation, tactics, imagination and creativity.

The threads across several chapters of embodied socialised subjects practising socially is pursued by Crang (Chapter 17). In a consideration of photography in historic site-visiting he argues that people practice leisure/tourism using objects like photographs and the activity of photographing, to articulate friendship, sociality and embodiment. Photography is used according to well acknowledged principles and learnt skill. However, rather than 'taking the picture' being understood as another example of the detached gaze, Crang 'sees' this as an often embodied and social event in itself. Moreover, the gathered image is taken home, passed around, worked, practised, and memories and knowledge reworked in a new embodied way, used as an object of friendship.

Whilst photographs can be passed round, plants can be swapped amongst social gardeners: taking photographs and working the ground amount to expressive practices. It may be that leisure/tourism practices generally have a shared materiality alerted in metaphor. Using empirical research on allotment holding and of rally caravanning Crouch (Chapter 18) discovers the significance of being surrounded by space in what we do, and in a way that works uneasily on strict demarcations of gender, but nonetheless acknowledges them. The two examples make very different sense of space, and perhaps surprisingly caravanning is observed as an expressive embodiment too, in a way that unsettles prevailing thinking towards leisure/tourism and its asserted categories (Cloke and Perkins 1998). This work brings together several themes of this book: aesthetic production, ritual and the embodied subject embedded in social relationships. Together, unevenly, this complexity of practice produces a participatory lay knowledge.

In a critical investigation of mid-century garden design Groening has tracked the use of leisure in representing nationalism, in that case, Nazism (Groening and Wolschke-Bulmahn 1997). In a very different approach in this volume, Niels Kayser Nielsen explores potentialities of a benign sense of nation in the way in which people practise leisure/tourism (Chapter 19). Indeed, leisure/tourism becomes a means of practising ideas and values such as identity, home, time and the space of practice. In the practice, these ideas and values are explored and reworked using space as a 'working partner', in which we move, do, where identity and self are processes. This chapter reopens and reworks many of the themes that cross this volume.

Concluding remarks

The chapters in this book are not subsectioned because their themes intertwine throughout, and this inter-connectivity is a key claim this book makes. Lay

11

geographical knowledge emerges from these essays as a process in which the subject actively plays an imaginative, reflexive role, not detached but semi-attached, socialised, crowded with contexts. The resulting knowledge in process resembles a patina and kaleidoscope rather than a perspective with horizon, a series of mutually inflected and fluid images rather than a map. Even a reflexive, 'inaccurate' map hardly resembles knowledge of this kind (Crouch and Matless 1996), even if cast as a momentary reflection, and overlays inappropriately formalise this knowledge. Reflexivity and liminality are important parts of this practice. Liminality represents less a distinct, temporal break than a chance to refigure, replay, remix other everyday spaces, representations, ideas and lives. This is, however, not limitless and works, often uneasily, within socialised constraints. The subject bends, turns, lifts and moves in often awkward ways that do not participate in a framing of space, but in a complexity of multi-sensual surfaces that the embodied subject reaches or finds in proximity and makes sense imaginatively. This combination contains meanings of landscapes, fragments, spaces, whole and abstract places, abstractions of the city and the country, street, nation, gender, ethnicity, class, valley, arena and field, through which human feelings, love, care and their opposites may be refracted. The subject mixes this with recalled spaces of different temporality.

Space and place tend to be used in two different relations in these chapters. Rather than impose one reading of this relation, this note will suffice. The words stand as one, an objective structure or context, the other that context practised by human subjects. The actual meaning of each is explicit in each chapter, although different authors use them in different sides of the relation.

The several themes that are intertwined through these chapters demonstrate fertile ground for further research. These chapters open a number of debates, directing developing concepts to tourism/leisure and enlarge its conceptual working. Extending insights across a range of more and less commodified leisure spaces and practices will test the limits of interpretation. Indeed, too often investigations of consumption have turned out to be investigations of the production of consumption. Enlarging the qualitative and ethnographic investigation of what people do, and make sense of, in leisure practices will improve the critical texture of understanding. There is a much needed extension of practices, spaces and knowledges towards a greater understanding of their social distinctiveness and relativity – yet, we argue, not to efface these dimensions of the imaginative embodied self.

There are also insights for connecting these conceptual debates through case studies to the future of such spaces and practices, often using the knowledge that proceeds with them. The development of policy towards leisure/tourism has been over-focused upon the market, commodification and its results. As such it becomes self-fulfilling, proving its ground and justifying more, measuring only its own results on its own criteria (Crouch and Tomlinson 1994). In many cases we argue that the results are much more complex, nuanced than is familiarly asserted. Three arenas of research that connect conceptual enquiry with policy concerns may be suggested here. Leisure/tourism remain important in terms of mental, physical and emotional health, but our understanding of how this works remains limited, although the opportunities for discovery and refiguring of the self are evidently important. Leisure/tourism practices can be significant in friendship, community-building, empowerment and identity as enjoyment.

Practices and their knowledge offer rich resources for popular rather than elite or formal care for the environment. This may be exemplified in terms of communicating understanding. Myers and Macnaghten have argued the limitations of the language used to communicate 'sustainability' because it does not work from popular language in everyday use (1998). The conceptual and applied use of commodity, consumption and product [sic] mystify what leisure/tourism in contemporary society mean. Working from lay practices and knowledge and familiar spaces, sustainability emerges as something much more accessible.

Understanding 'Heritage' in terms of lay geography directs attention to the tourist/leisure representation and practice of everyday life. This is less a debate about authenticity than about empowerment and identity. Surveys that work from preferences of what is promoted inevitably pre-limit their intelligence. Heritage development and empowerment/identity are issues of interest and pursuit amongst subjects on 'both sides'. Unsettling heritage by realising and articulating contemporary heritage in process is a particular component of the project Representing Weardale (Crouch and Grassick 1999).

The complexity of spatial practice is understood as embodied socialised subjects living amongst particular social relations. These different components are engaged unevenly, sometimes unwillingly, and together subject to an ongoing reflexivity. As this author discusses elsewhere, the partially abstracted landscapes – or figuratively represented lay geographies – of some of the mid-twentieth century painters resemble more closely the knowledge made of, and in, spatial practice (Crouch and Toogood 1999). Whilst not immediately accessible to the eye gazing for a clearly observed and still moment, such more fractured, multiple geographical knowledge as represented on the cover of this book is likely to be closer to our lived practice.

References

Augé M. (1995) *Non-Places: introduction to an anthropology of super-modernity*, Verso: London.
Bachelard G. (1994) *The Poetics of Space*, Beacon Press: Massachusetts.
Bale J. (1993) *Landscapes of Sport*, Leicester University Press: Leicester.
Bauman Z. (1993) *Postmodern Ethics*, Oxford: Blackwell.

Bishop K. and Hoggett P. (1986) *Organising around Enthusiasms: patterns of mutual aid in leisure*, Comedia Publishing Group: London.

Bourdieu P. (1984) *Distinction: a social critique of the judgement of taste* (trans R. Nice), Routledge: London.

Bourdieu P. (1990) 'From rules to strategies: an interview with Pierre Bourdieu', *Cultural Anthropology* 1(1): 110–120.

Clark J. and Critcher C. (1985) *The Devil makes Work: leisure in capitalist Britain*, MacMillan: London.

Clarke G. *et al.* (1994) *Leisure Landscapes*, Council for the Preservation of Rural England/Lancaster University: London.

Cloke P. and Perkins H. (1998) ' "Cracking the canyon with the awesome foursome": representations of adventure tourism in New Zealand', *Environment and Planning D: Society and Space* 16: 185–218.

Crang P. (1994) 'It's showtime: on the workplace geographies of display in a restaurant in southeast England', *Environment and Planning D: Society and Space* 12: 675–704.

Crang P. (1996) 'Popular geographies', guest editorial, *Environment and Planning D: Society and Space* 14: 1–3.

Crouch D. (1992) 'Popular culture and what we make of the rural', *Journal of Rural Studies* 8(3): 229–240.

Crouch D. (1994) *The Popular Culture of City Parks*, Comedia Demos Working Paper 9: London.

Crouch D. (1998) 'The street in the making of popular geographical knowledge', in N. Fyfe (ed.) *Images of the Street*, Routledge: London.

Crouch D. and Grassick R. (1999) *People of the Hills*, Amber/Side: Newcastle.

Crouch and Matless (1996) 'Refiguring geography: the parish maps of common ground', *Transactions of the Institute of British Geographers* 2(1): 236–255.

Crouch D. and Tomlinson A. (1994) 'Leisure, space and lifestyle: modernity, post-modernity and identity in self-generated leisure', in I. Henry (ed.), *Modernity, Postmodernity and Identity*, Leisure Studies Association: Brighton.

Crouch D. and Toogood M. (1999) 'Everyday abstraction: geographical knowledge in the art of Peter Lanyon', *Ecumene* 6, 1: 72–89.

de Certeau M. (1984) *The Practice of Everyday Life*, University of California Press: Berkeley.

Finnegan R. (1989) *The Hidden Musicians: music making in an English town*, Open University Press: Buckingham.

Finnegan R. (1997) 'Music, performance, enactment', in G. Mackay (ed.) *Consumption and Everyday Life*, Sage: London.

Fyfe N. (ed.) (1998) *Images of the Street*, Routledge: London.

Game A. (1991) *Undoing the Social: towards a deconstructive sociology*, Open University Press: Buckingham.

Giddens A. (1991) *Modernity and Self-Identity*, Polity Press: Cambridge.

Gorz A. (1984) *Paths to Paradise*, Pluto Press: London.

Gregson N. and Crewe L. (1997) 'The bargain, the knowledge and the spectacle: making sense of consumption in the space of the car-boot fair', *Environment and Planning D: Society and Space* 15: 87–112.

Groening G. and Wolschke-Bulmahn J. (1997) *Grune Biographien*, Hannover Patzer: Berlin.

Harré R. (1996) *The Discursive Mind*, Blackwell: Oxford.

Jackson P. and Thrift N. (1995) 'Geographies of consumption', in D. Miller (ed.), *Acknowledging Consumption*, Routledge: London.

Kayser Nielsen N. (1995) 'The stadium and the city', in J. Bale (ed.) *The Stadium in the City*, Keele University Press: Keele.

Kristeva J. (1996) *The Portable Kristeva*, Columbia University Press: New York.

Lash S. and Urry J. (1994) *Economies of Signs and Space*, Sage: London.

Ley D. and Olds K. (1988) 'Landscape as spectacle: world's fairs and the culture of heroic consumption', *Environment and Planning D: Society and Space* 6: 191–212.

Macnaghten P. and Urry J. (1998) *Contested Nature*, Sage: London.

Maffesoli M. (1996) *The Time of Tribes*, Sage: London.

Massey D. (1993) 'Power-geometry and a progressive sense of place', in J. Bird *et al.* (eds), *Mapping the Futures: local cultures, global change*, Routledge: London.

Matless D. (1995) 'The art of right living: landscape and citizenship 1918–1939', in S. Pile and N. Thrift (eds) *Mapping the Subject*, Routledge: London.

Merleau-Ponty M. (1962) *The Phenomenology of Perception* (trans. R. Nice), Routledge: London.

Miller D. *et al.* (1998) *Shopping, Space and Identity*, Routledge: London.

Moorhouse R. (1992) *Driving Ambitions*, Manchester University Press: Manchester.

Myers G. and Macnaghten P. (1998) 'Rhetoric of environmental sustainability', *Environment and Planning D: Society and Space* 30: 333–353.

Pile S. and Thrift N. (1995) *Mapping the Subject*, Routledge: London.

Radley A. (1990) 'Artefacts, memory and a sense of the past', in D. Middleton and D. Edwards (eds), *Collective Remembering*, Sage: London.

Radley A. (1995) 'The elusory body and social constructionist theory', *Body and Society* 1(2): 3–24.

Radley A. (1996) 'The triumph of narrative: a reply to Arthur Frank', *Body and Society* 3(3) 93–101.

Reville G. and Watkins C. (1996) 'Educated access: interpreting Forestry Commission Forest Park Guides', in C. Watkins (ed.), *Rights of Way: policy, culture, management*, Cassell: London.

Rojek C. (1993) *Ways of Escape*, Routledge: London.

Rojek C. (1995) *De-centring Leisure*, Sage: London.

Selwyn T. (1996) *The Tourist Image*, Wiley: London.

Shields R. (1992) *Lifestyle Shopping*, Routledge: London.

Shoard M. (1997) *The Land is Ours*, Palladin: London.

Shotter J. (1993) *Cultural Politics in Everyday Life: social constructionism, rhetoric and knowing of the third kind*, Open University Press: Buckingham.

Thrift N. (1997) 'The still point: resistance, expressive embodiment and dance', in S. Pile and M. Keith (eds), *Geographies of Resistance*, Routledge: London.

Urry J. (1990) *The Tourist Gaze*, Sage: London.

Urry J. (1995) *Consuming Places*, Routledge: London.

Warren S. (1993) 'This heaven gives me migraines: the problems and promise of land-scapes of leisure', in Duncan J. and Ley D. (eds), *Place/Culture/Representation*, Routledge: London.

Young I. M. (1990) *Throwing Like a Girl and Other Essays in Feminist Philosophy and Social Theory*, Indiana University Press: Indiana.

Zukin S. (1995) *The Cultures of Cities*, Blackwell: Oxford.

2

THE MYTHO-POETIC IN NORTHERN TRAVEL

Inger Birkeland

Impressions from the edge of the world

Standing at the cliffs at North Cape I saw the same landscape as other visitors have seen for centuries, and simultaneously I saw a different landscape. Looking down I saw the cliffs fall straight down into the sea. Looking north I saw a horizon where the sky merged with the sea. I had travelled as far north as it was possible to travel on land. It was not possible to travel farther north without crossing a very material border, the cliffs which marked the boundary between earth and sea. I had reached the north point of Europe's mainland, the edge of the world, known for centuries by sailors, explorers, traders, travellers and tourists. When I had the cliffs of North Cape under my feet, I had a strange feeling of being two places simultaneously.

In the evening we were waiting for the deep red midnight sun. I was alone but didn't feel lonely. We were many who shared the act of waiting for the midnight sun. It was easy to see the pleasure in the faces of the many travellers outdoors at the cliffs. I saw intended and desired looking at the sun; waiting for the sun to descend in the horizon, waiting for the sun to rise again. Being there was both an individual and a social event. Most of us arrived there as strangers to each other, but the act of watching the midnight sun seemed to transform strangers to family and friends at an unordinary level. Even if we were strangers to each other, there was a mutual seeing of the same deep red sun. I heard many unfamiliar words uttered in other mother tongues, and as more and more visitors arrived at the cliffs, I felt like I was walking in a multicultural, multicoloured city. But the fear I sometimes feel in the city was absent. Even if we could not speak each other's languages very well, we communicated easily. Someone spoke of what they saw, the beautiful colours played out on the sea and in the sun that was partly covered by clouds low in the horizon. We saw it and we understood it. The words uttered were the uncomplicated, the kind of words that sound trivial outside the space of there and then. But they were not trivial, rather they represented another way of creating meaning out of the meaningless, Order out of Chaos, light out of darkness. Our mutual looking at the sun and our waiting for the sun to descend in the horizon made us very human there at North Cape, an edge of the modern world. These acts of waiting and watching were uttered in a silent language, as the most simple acts in the world, and simultaneously the most complicated.

When the midnight sun reached its lowest position in the horizon, my curiosity rose. I felt that if the sun would rise again it would be a complete mystery. The sun was out there in front of me, and I started to wonder what could be found beyond the edge of the cliffs. I knew the ground that I trod, a dusty ground covered with gravel and sand, but not the quietly floating dark sea and the cool blue sky and the red sun. I could not know what was beyond the cliffs – or could I? Could it be that the border I saw in the cliffs represents the threshold between the conscious and the unconscious, the Great Border between Logos and Mythos? Might it be that the crossing of this border might represent a departure to the land of Mythos? If I was to travel farther, I knew that I had to leave the land of Logos and move into a new country of meaning, the time-space of myth and mythmaking.

When the sun did rise, it dawned on me that ordinary ways of using language were insufficient to describe the meanings of North Cape. I suddenly knew that I ought to use words sharp as the line of black ink on a white page, analytical tools that can cut through the world in rational and violent ways. As I was standing at the cliffs watching the midnight sun I discovered North Cape as the birthplace for a transformation of my way of being and thinking, which in the moment seemed non-intentional and non-rational but nevertheless was inscribed in some kind of language. Looking at the sun rise again to a new day made me realise this impossible and necessary border in the world as a birth place of the poetic creation of myth.

The sudden revelation stemming from watching the sun rise to a new day turned my arrival at North Cape into a new journey. As I started to transform my impressions of the edge of the world into fragmented short stories, small snapshots of meaning emerged. These thoughts moved as the horizon moved and the eyes drifted together with the bodily movements. New impressions were caught in my walking in-between other visitors there on the carpet of gravel and sand, where we were navigating our way around each other and preparing for the new day. It was just like being in the midst of living a life on the road between birth and death, between sleeping at one-night-only hotels and eating salmon dinners at roadside restaurants. I suddenly felt like an eternal wanderer, a modern nomad, an inhabitant in the land of Mythos, and I knew that I would continue to cross borders and transform the violent language of Logos to a

Plate 1 The North Cape sunrise depicted on a postcard, 1933

language of myths, telling stories of birth and renewal. And I felt like an explorer and discoverer travelling in the history of eternal space and in the memories of the present. Arriving at North Cape was just like being at home in a strange place, and being a stranger coming home. I started to feel that the rising sun turned my new day into a 'long poem of walking', as Michel de Certeau has so beautifully put it (1984: 191).

Northern travel

I have introduced the theme for this essay by telling a story which is intended to evoke particular ways of seeing North Cape[1]. The impressions from the edge of the world described above indicate that the journey to the north opens up a possibility for a change towards something good, beautiful or true. Generally travel can be characterised as an exemplary case of personal change. The attraction of travel is related to the experience of having been somewhere and having seen something with one's own eyes because of a curiosity to learn and understand. Travel is thus a process that opens up understanding and knowledge for individual human beings, always historically and geographically specific. In this essay, I will discuss some aspects of the change in the journey to the north in modernity in terms of an engagement with the mytho-poetic, which here refers to a poetic creation of myth.

The interest in the poetic creation of myth is inspired by Michel de Certeau (1984: 105), who relates the mytho-poetic to practices and activities which might articulate a 'second, poetic geography' on top of the practical time-space of modernity. I understand modernity as the only time in history where human beings 'attempt to live without religion' (Kristeva 1986a: 208). In addition, I also see modernity as a space, a secularised space constituted by a lack of place (de Certeau 1984: 103). Modernity is characterised by a production of abstract and anonymous places where the 'body of legends' specific for pre-modern places have been left out, according to de Certeau (1984: 106–107). Against this back-ground, de Certeau argues that storytelling plays an important role for a restoration of place in modernity. The reason is that travel and storytelling share in being ways of moving into otherness and discovery of an other place (1984: 109). In this chapter I will discuss the arrival at North Cape as both a discovery and a restoration of place by a poetic creation of myth.

This discussion will take its outset from a certain perspective, which is the perspective of the woman as the foremost model of a traveller and storyteller. To me the restoration of place in modernity is related to mythmaking and to mythic language as the language of the woman traveller. I am inspired here by Julia Kristeva's idea about the woman as the foremost model of the stranger, because of woman's position as a stranger in the symbolic order (Kristeva 1984, 1991, 1986b, Fürst 1995, 1998, Smith 1996). I see the woman as a traveller in the symbolic order, in other words in the land of Logos, whose particular posi-tion represents a possibility to speak about other things, tell different stories and raise novel questions. In the following I will contextualise this argument as it relates to the impressions of seeing the midnight sun at North Cape in three ways. First, historical travel to North Cape will be represented as the outcome of a masculine quest for knowledge and truths, and with North Cape as a projec-tion of masculine images of verticality and horizontality. Second, the arrival at North Cape will be discussed from the point of view of a woman traveller. The arrival at North Cape for the travelling woman represents both an arrival in the symbolic order and a transformation of this arrival into a new journey, a search for a symbolic order more appropriate for the woman traveller. Finally, I will discuss this transformational work by drawing upon the meanings of the sun in ancient mythology. The arrival at North Cape will be redescribed by focusing on creation myths in northern mythology, where the sun represents a deity that expresses feminine and fertility-oriented creation.

The quest for the North Cape

Travelling north has always been considered risky. Nevertheless, travel to the cold, dark and inhospitable territories of the north must also have been attractive for those departing. In the centuries after the Renaissance the coasts and interior of the northern parts of mainland Europe were gradually mapped and colonised, both materially and symbolically, by the Scandinavian countries with the aid of European explorers and traders. Thus, the northern territories, formerly *terra incognita* to the rapidly expanding nations of Europe, were discovered and named. For example, North Cape was first marked on a map dated 1553 made by the navigator Richard Chancellor, participant on the Willoughby expedition which set out to find the northern passage to China (Skavhaug 1990; Jakobsen 1997). The expedition failed. North Cape was thus discovered, named and produced as a place for certain historical purposes and projects, and the historical development of North Cape as a travel destination grew out of the early exploratory travel.

One of the earliest travellers to North Cape that we know of was the Italian Francesco Negri, who visited North Cape twice, first in 1663 (Skavhaug 1990; Jakobsen 1997). He described the experience of visiting North Cape as being at an utmost point and at the edge of the world. Negri experienced that he had seen all that was to be seen, his curiosity ending as his journey came to an end. North Cape was the edge of the world to him because there was no place further north populated by human beings. After seeing this, he wrote in his diary, he could return happily back home. To most people in the seventeenth century, places like North Cape were invisible, unseen and unknown, located outside of first-hand experience.

Much has changed since Francesco Negri's first journey to North Cape in 1663. Commercial and recreational travel developed with the advent of modernity. Institutionalisation and commodification of leisure, coupled with new transport technologies, were important for the development of North Cape as a holiday destination for the privileged members of the European population. As early as 1866 Thomas Cook & Son started planning tours to Scandinavia and North Cape. In the report on the first trip in *Cook's Excursionist and Tourist Adviser* (9 December 1875) the focus is not only on the practical problems. What is striking is that North Cape was made the goal ('the extreme point') of the tour. The climax was to see the sun at midnight ('the midnight sun'). At the turn of the century there was already a large amount of travel literature on North Cape, and among this literature there are travel handbooks and more or less autobiographical travel stories. One of the books is a description of the people and nature in northern Norway written by Hans Reusch (1895). Reusch's description of North Cape is striking:

> On the voyage from Hammerfest eastwards the steamer took a turn out to North Cape. This mountain which is flat on top and which has steep walls, is so often described that I do not need to mention it

further. The fame of North Cape is a peculiar proof of the role the aesthetic plays for human beings. North Cape is in fact, as has been shown a long time ago, not the northernmost point of our continent. Knivskjærodden just west of North Cape is stretching farther north in the sea, but since this point is low-lying and not much to look at, nobody takes notice of the correct observations and measures. North Cape attracts everybody both because of its looks and its telling name. North Cape is poetry, Knivskjærodden is prose. The latter, prose, we have enough of in daily life, therefore we also have seen so many toasts of champagne at North Cape, but not one at the other point with the long name.

(Reusch, 1895: 16, my translation)

In Reusch's description I find a differentiation between aesthetic and non-aesthetic experience, and an explicit celebration of the aesthetic qualities of North Cape. This aesthetic experience is illustrated by Reusch's distinction between North Cape and the neighbouring cliff, Knivskjærodden, in the differ-ence between the poetic and the prosaic; North Cape is poetry while Knivskjærodden is prose. This way of seeing North Cape might reflect the nine-teenth century romanticist view of nature, which stresses feeling, emotion and imagination, and where the sublime and the picturesque are major aesthetic cat-egories for the experience of nature (Jasen 1995: 7). Seeing nature as sublime is particularly important, Jasen says, because it deals with experiences which sometimes create an enhanced state of being, wonder and beauty, and sometimes anxiety and terror, which leave people speechless with awe. Views of nature as wild, extreme, dangerous, magnificent and overwhelmingly beautiful are found in early modern travel in different areas of the northern hemisphere, and Jasen says that such feelings can be understood as substitutes for religious experience, which in the nineteenth century were transferred to the area of tourism and leisure (Jasen: 7).

At North Cape *the sublime* is an appropriate expression for seeing the sun during the night, when the sun is supposed to travel below the horizon, for seeing the form of the land, which represents archaic, arctic plains and dark bluish-green cliffs falling one thousand feet down into the sea. Simple lines of verticality and horizontality are represented by the sharp border between earth and sea, visible for all visitors arriving at North Cape. At the turn of the century, visitors arrived at North Cape from the sea. Steamers anchored in a nearby bay (*Hornvika*), where visitors were set on shore in order to ascend to North Cape on foot, using one's own body by climbing one thousand feet to the summit of the cliffs (Jakobsen 1997). In other words, visitors saw and experienced a dramatic landscape, magnificent and dangerous simultaneously.

Plate 2 Near the North Cape (photo: Inger Birkeland)

The development of North Cape travel in this century has turned North Cape into a place of consumption for thousands of travellers arriving from all parts of the world. The development in new transport technologies in this century, in particular the automobile, have contributed to a major growth in the number of travellers to North Cape (Jakobsen 1997). This has however created a major change in the experience of North Cape as well. In 1956 the mainland road to North Cape was finished, and Jakobsen (1997) has argued that easing the access changed the frame for the experience of North Cape to one less elevated and less impressive. I will suggest that the change in access to North Cape also changed the experience of the verticality and horizontality at North Cape, where the vertical element experienced by climbing the cliffs was lost in favour of a flat, horizontal framing. Verticality is today experienced visually, and only from above. However, the representations of North Cape in advertising and sales materials etc. still reproduce the main frame of the striking profile of the vertical cliffs dividing horizontal plains from arctic ocean.

Historical travel as other quests for knowledge may be interpreted in terms of discourses of masculinism (see Pratt 1988, Mills 1991, Rose 1993, Blunt 1994, Blunt and Rose 1995, Birkeland 1997). Masculinism as I see it refers to male notions of travel and knowledge on the background of a connection between masculine subjectivity and claims on knowledge (Rose 1994: 5). Masculinism is also articulated in present tourism and leisure, for example in production, inter-pretation and consumption of heritage (Edensor and Kothari 1994). I would like to focus on an interpretation of the visual framing of North Cape as an articu-lation of masculinism. The verticality and horizontality of North Cape may be interpreted in terms of an abstract logic of visualisation (Lefebvre 1991). Henri Lefebvre argues that verticality and horizontality always have been concrete spatial expressions for great powers of war, and that this is a principle which is not only related to a logic of visualisation but to a phallic logic. This logic creates simplification and reduction of meaning through horizontal and vertical lines in the production of modern places, i.e. in towers and skyscrapers (Lefebvre: 318). This form of constraint or repression is at North Cape identifiable in a framing of particular viewpoints and perspectives. Such reduction is nothing more than an arrogance and a will to power, an alliance between Ego and Phallus, Lefebvre says (Lefebvre: 261). However, Lefebvre also sees the articulation of this phallic logic in the area of leisure and tourism as a potential for critique. He says that leisure space may take a particular role in the restoration of meaning, and that 'the feminine principle' is important in this restoration (Lefebvre: 248). Lefebvre even indicates that 'in and through the space of leisure, a pedagogy of space and time is beginning to take shape' (Lefebvre: 384).

The woman traveller as myth-maker

As a starting point for this pedagogy of space and time I will focus on a reading of the arrival at North Cape from a psychoanalytic point of view. The arrival at North Cape will be read as an arrival in the symbolic world from the stand-point of the woman traveller. This discussion will be based on Julia Kristeva's (1984) ideas of the heterogeneous subject and poetic language.

In *Revolution in Poetic Language* Kristeva develops an understanding of the human being as a subject in process/on trial, and to me that is a travelling subject (see also Smith 1996). The subject in process/on trial is also named the hetero-geneous subject, which is constituted by semiotic and symbolic dispositions. The symbolic is here synonymous with the Law of the Father in Lacan's psychoana-lytic theory. The semiotic, on the other hand, is described as 'chora' (Kristeva 1984: 25). It seems that chora is associated with an absolute place that comes before the entry in symbolic language. Chora develops through spatial intuition and has an analogue in vocal or kinetic rhythm, seasonality and repetition. Chora is also associated with the unconscious and with the maternal first place. Chora is the place of mother-child relations where '[t]he mother's body acts with the child's as a sort of socio-natural continuum' (Kristeva 1986c: 148). This means that the heterogeneous subject travels in both symbolic and semiotic language, which we also might think of as journeys in the lands of Logos and Mythos.

24

Plate 3 Land meets sea at the North Cape (photo: Inger Birkeland)

The heterogeneous subject realises poetic language. Poetic language is not solely based on language's poetic function. It is rather a possibility inherent in all meaning production, as all types of speech or language acts can be realisations of the poetic, Kristeva says. Poetic language parallels the unconscious and communicates *jouissance* and regression, or what Kristeva terms the 'revolution in poetic language' (Kristeva 1984). In Kristeva's inherently spatial language poetic language thus is understood as the process which makes the feminine place of chora speak or travel. Poetic language is not possible to decide, or capture and locate. It is described as a passage in and through language to the outer boundaries of the subject and society (Kristeva 1984: 17). This is a very spatialised way of representing language and meaning, where meaning construction is seen as a process of passage, and the meaning obtained as an arrival at the borders.

The notions of the travelling subject and poetic language are relevant for an understanding of the journey to North Cape. The creation of poetic language as an ongoing, unbounded and generating process can be interpreted as analogue to the process of travel, and where the woman traveller represents the semiotic chora. We might see this woman's journey to North Cape as a departure from home with a destination at North Cape as an entry in the symbolic world. As Kristeva argues (1996), the arrival in language is experienced differently for men and women. For a woman traveller the journey starts with a departure from home with the separation from the mother, to the discovery of the father, associated with language, the Phallus and the symbolic world. Kristeva (1996) argues that the symbolic world is experienced as a strange world to women, it is like an arrival at a strange place. The world may thus be experienced as illusory, i.e. not unreal, but something strange, not untrue, but a play or a game.

Thus we may see the woman as the foremost model of both the stranger and the traveller in society. Kristeva suggests that this position provides the woman with possibilities for developing an ethics of difference and an ethics of love connected to motherhood, feminine creation and imagination (Kristeva 1986b, 1987, Fürst 1998: 193–194). Kristeva's ethics is based on psychoanalytic practice, and Fürst (1998) shows that this also implies a politics of difference. I would like to suggest a possibility for an engagement in a feminine ethics that relates to a poetic creation of myth as a feminine form of creation, and in particular an engagement with creation myths.

Human beings have always created myths of a first place as a sort of mysterious origin of nature and culture, and therefore creation myths are deeply embedded in religious belief. Creation myths concern the fundamental question: Who created the world? Where do I come from? Marie Louise Von Franz argues that whenever human beings have arrived in a new and strange place, they have tended to project images that express the basic myths of creation in their culture (Von Franz 1995: 2). Myths are fantastic creations of human beings. They are constituted in much the same way as dreams are constituted, as another reality, in images, in art, literature or in other ways (Lauter 1984). Myths are realised in particular situations, i.e. in initiation, marriage, birth, in the change of the four seasons of the year, and in the change between night and day. The yearly rhythm between work and holiday may be another such situation which calls for the projection of mythical images. Creation myths have in particular arisen from projected images of the familiar in an unfamiliar place, and are performed and repeated under certain conditions. When medieval peoples for example set foot in a new land they repeated creation myths as a manifestation that the land had not formerly existed. After arrival in the new land, it entered their consciousness, and only at that point they could create the new land, and settle. Thus creation is about consciousness that is awakening, a *consciousness-of-the-world*.

Different creation myths have existed side by side in different religious beliefs. Julia Kristeva (1986c) says that once upon a time patriarchal monotheism followed after a more material or fertility-oriented religion. Patriarchal monotheism thus won a war between the sexes over the creation of the world (1986c). We know that Judeo-Christian religion is monotheistic, and that the creator is represented as a singular God, a male creator. Thus we hear stories of a man who created the world:

> by dividing light from darkness, the waters of the heavens from the waters of the earth, the earth from the seas, the creatures of the water from the creatures of the air, the animals each according to their kind and man (in His own image) from himself. It's also by division that He places the opposite each other: man and woman.
>
> (Kristeva 1986c: 139–140)

The creation myth of Judeo-Christian religion tells us that the world is a place created by a singular God. Luce Irigaray (1993) argues that this male creator has injected himself into time, consciousness and the enlightenment of the daily sunshine, while he has closed himself off from the feminine place where orientation starts from, the great abyss of darkness and night. Irigaray argues that this closure over time has resulted in representations of woman as a place, as a thing or an object. Woman as place became an envelope or a container for the man and a maternal container for his child. This space he does not know in other ways than through repression or submission to everything that woman is – including the semiotic chora, and the earth as Mother Earth. Because once this creator was enveloped in a container, the container space of woman is threatening. The consequence is that woman has become a forgotten place in the habitable regions of modernity, she has become synonymous for the place of lack. This is what is presented to us as the problem, according to Irigaray (1993). Julia Kristeva's (1987) diagnosis is that we do not have a discourse on motherhood and feminine creation in modernity.

How does this relate to the arrival at North Cape for the travelling woman? Might the arrival at North Cape not only represent a discovery of symbolic language, but also a rediscovery of semiotic language? It is possible to interpret the change in the journey to the north as a discovery of mythic language. There are elements in the story of the impressions from North Cape connected to seeing the midnight sun that leads to such an interpretation. In addition, in the mythologies of northern peoples there are fragments of meaning which bring forth an interesting perspective on the feminine as a separate form of symbolic order. Northern mythology explicitly expresses polytheism, and creation and creativity is here not only understood as an expression of the masculine principle but as a particular feminine and life-affirming power. In the following I will try to describe how this can be so in relation to North Cape travel. I ask: what happens when the woman traveller starts telling her stories about seeing the midnight sun at North Cape?

To see the midnight sun: participation in feminine creation

When the sun ascends in the horizon at North Cape in the middle of the night it seems that silence and astonishment is the first reaction. However, to feel estranged or silenced by one language may also represent a possibility to speak another language. Watching the sun at midnight is something that people in these regions of the world have done as long as there's been people living there. The farthest north of the European continent has for thousands of years been popu-lated by the ancestors of the semi-nomadic Sami people, Norsemen, Finns and others who have maintained and developed their life-modes in a very marginal environment. Seeing the midnight sun is possible at this latitude a few weeks during the summer because of the earth's changing position in relation to the sun over the year. If one lives above the Polar Circle, which marks the border for the appearance of the midnight sun in the horizon, one has to relate to the particular rhythm of the seasonal changes over the year. In these areas the rhythm of day and night is extended and blended with the cycle of the year. This temporal change seems like a movement between one long day of a summer and one long night of a winter. Everything that makes a difference becomes polarised in terms of day and night, summer and winter, warmth and cold, light and darkness. The fight for life is a real fight against the colonisation of life by the long, cold and dark winter.

Habitation of the north is risky, because of the struggle against darkness, cold-ness and winter, as is indicated in the ancient myth of *Ultima Thule*, which is a name for the farthest north and the limit of any journey, the edge of the north-ernmost world (Jakobsen 1997). One has to be careful not to 'travel north and under', a saying still goes in Norway. However, when the summer comes the north has transformed itself to a heavenly place of eternal light, embracing both day and night into a continuous day. People living in the more temperate regions do not experience these rhythms between light and darkness in the same polarised way. These seasonal changes give an additional dimension to life which can only be experienced by travelling north for those who do not have the north as their home. The sun thus stands in the centre of this eternal struggle between light and dark, very important in the mythologies of northern peoples.

European mythology provides us with many myths of sun, which are connected to a range of different and ambiguous meanings (Green 1991). On the one hand the sun is associated with life, creation and healing, fertility and light, heroic journeys, conquering of death, but on the other hand the sun is also associated with destruction and death, power and control, the eye of the universe (ibid.). Green (1991: 136) says that the mythic images of the sun reveal 'a conflict between good and evil, light and darkness, life and death, summer and winter, positive and negative'. Green does not explicitly include the dimensions of feminine and masculine in this double image of the sun. Eliade (1957) argues that the sun in classic mythology has an opposite meaning of the moon. The sun was associated with intelligence, consciousness, power, autonomy, authority, creativity and knowledge, in classically masculine characteristics. The moon on the other hand was associated with feminine qualities: birth, death, becoming, cosmic darkness, water, madness, the unconscious. In classic mythology the meaning of the sun was closely associated with hero myths, where the sun in the figure of the hero fights and wins over darkness. Eliade argues that over time the heroic struggle between light and darkness was transformed (1957: 158). The sun became associated with intelligence and consciousness without its other, the meaning of darkness. Eliade suggests that this produced rationalisation and a cultivation of consciousness to such a degree that the place of darkness in the mythic order was lost. The sun as deity was replaced with pure ideas (ibid.).

Green (1991: 107) points out that the sun does not have identical meaning in the mythologies around the world. Worship of the sun has for example been absent in the hotter regions of the world, she says, because the sun in these areas was associated with destruction of life rather than creation of life. Worship of the sun as a provider of life has been particularly important in the northern regions of the world, according to Green (ibid.). An association between the sun and fertility, creation, and conqueror of death, can be traced in the northern areas of Europe. This opens up interesting perspectives. Does this suggest a possible association between the sun and feminine aspects of creation?

Goddess worship was important in ancient Greece, as it is in all maternal or fertility-oriented religions, such as Norse and Sami. We know that people in the northern parts of Europe in ancient times worshipped both paternal and maternal divinities. In Norse religion the world of gods contained a heterogeneous and nuanced idea of the divine, which was much more sophisticated than what is found in Judeo-Christian religion (Steinsland 1997). According to May-Lisbeth Myrhaug (1997), Sami religion has an even more developed goddess worship than Norse religion. The Sami worldview holds that the universe is filled with deities both above the heaven, in the heaven, on the earth and underneath the surface of the earth (Myrhaug: 27). This is a complex worldview where gods and goddesses create a unity or balance between 'the feminine' and 'the masculine' principle. An important creation myth is about the sun. In Sami religion the sun has a feminine figuration (ibid.: 83). The sun (Beive) is called queen in heaven, she has power and influence over life on earth, and is expressed by an image where four rays are spreading out from a centre. These rays symbolise the four earth corners and illustrate the placement of all living things in relation to the sun. In the moment of creation the soul was carried to each corner until given a body by the mother goddess (*Máttaráhkká*), in the final fourth corner.

The figuration of the sun as feminine in northern mythology offers a re-interpretation of seeing the sun at midnight. This opens up for seeing creation literally in a different light. Seeing the sun in the middle of the night is done at a time of the day when the moon, lunar myths and rhythms are sovereign. In lunar myths the principle of becoming and renewal is central. Seeing the sun descend and then ascend at midnight provides the sun deity with the dimension of becoming, in other words with the feminine principle consisting of the flow of life: life, survival and creativity. The midnight sun is thus an image of darkness as a source of light. Looking at the midnight sun is a strange reconcil-iation of the polarity of the sun and the moon, the midnight sun reconciled with the meaning of darkness in the mythic order. The restoration found in looking at the midnight sun at North Cape represents a new meaning of consciousness, a sudden revelation of the meaning of daylight. The midnight sun represents the becoming of feminine consciousness and the new morning of life, where the meaning of consciousness has included the feminine principle. To wake and to dawn to a new day at North Cape, to gain knowledge of a different kind, is the enlightenment we are given by seeing the sun at midnight.

Conclusion

In this chapter I have represented an experience of seeing the midnight sun at North Cape in a particular way. The arrival at North Cape has been represented as a discovery and transformation of experience by the help of creation myths. The arrival in the north has been rewritten as a movement, of reaching out for a restoration of the poetic imagination through mythmaking.

North Cape has been described as an isolation of a phallic logic, as a projec-tion of an image of the masculine principle. This phallic logic was identified as a fixation on the vertical and horizontal image of the North Cape profile, as an image of a language that cuts off the flow of meaning and refuses the rhythmic renewal of life. The masculine principle, symbolised by the cliffs dividing earth and sea at North Cape, represents an articulation of violent separation, creation of borders and compartments of meaning. However, this same border between earth and sea at North Cape was also represented as a name for a departure to another region of meaning, the land of Mythos. With the image of a travelling woman arriving at North Cape, the arrival in the symbolic order was trans-formed to a search for the language of Mythos and a creation of stories of life and renewal of life. The discovery of the meaning of mythmaking was made possible by experiencing the sun turn the night into a new day.

The interpretation of the northern sun stresses the feminine aspects of creation and creativity. The mythic creation of feminine consciousness described through the image of the midnight sun represents both a critical and a poetic project. The feminine sun tells us that there are new fragments of meaning waiting to be told in the in-betweens of the great journey of life. The new consciousness is a continuing and simultaneous process of movement and narration, as if consciousness was similar to walking, looking, taking a step ahead, searching the best path to follow. The change in the journey to the north is thus a discovery and a recovery of the dynamics of life, a discovery of the birthplace of poetic imagination and a recovery of place. The experience of arrival to North Cape makes it possible to transcend the loss of place characteristic in modernity by a creative invention of myths of creation where feminine creation and feminine consciousness are restored. The change in experience is thus a change towards opening up for consciousness of a feminine way of reaching and relating towards the world, which has relevance for both men and women.

Having been to the north, every journey south seems to be a second journey towards home. The departure for home may involve the employment of the new knowledge and represent a re-creation of the potential to restore the experience of having a place in the world. It may open up for other and more ambiguous understandings of life and creativity where darkness may be seen as a hiding place of light, a paradoxical source of novel meaning. The new understanding of knowing and being, where the rising sun was represented as a symbol of feminine creation, can be seen as a complicated transformation of individual experience, which creates a feeling of wonder that brings new and unfamiliar meaning into the familiar world of home, work and leisure. The new consciousness that may be brought home will hopefully have left the house of double binds and either/or thinking, the polarity of positions for or against, the lessons taught as Reason without Passion, Logos without Mythos. The new way of thinking and feeling will hopefully be on the side of motion and e-motion. The new-born home-comer might perhaps wander from the place of questions, start looking from the position of a pathfinder, an explorer of traces and tracks, myths and legends, and whisper to you from this mobile position: Who are you? Where are we going? Is this the right track?

Acknowledgements

I would very much like to thank audience and convenors of the session on Contemporary Leisure and Tourism at the RGS/IBG Conference in Exeter, January 1997, where an earlier version of this chapter was first presented. Thanks in particular to David Crouch, Luke Desforges and Jon May. Thanks to Arild Hovland, Elisabeth L'Orange Fürst, Gillian Rose, Kirsten Simonsen, Victoria Einagel and Martha Easton for comments and suggestions at different stages in the process of writing.

Notes

1   This chapter is based on my doctoral research on modern holiday travel to North Cape, Norway, in the 1990s, and in particular on participant observation at North Cape in July 1996.

References

Birkeland, I. (1997) Visuell erfaring som situert kunnskap. *Sosiologi i Dag* 27, 1: 73–89.

Blunt, A. (1994) *Travel, Gender and Imperialism. Mary Kingsley and West Africa,* London: The Guildford Press.

Blunt, A. and G. Rose (1994) *Writing Women and Space. Colonial and Postcolonial Geographies,* London: The Guildford Press.

Cook's Excursionist and Tourist Adviser, 9 December 1975.

de Certeau, M. (1984) *The Practice of Everyday Life,* Berkeley: University of California Press.

Edensor, T. and U. Kothari (1994) The masculinisation of Stirling's heritage, in Kinnaird, V. and D. Hall (eds) *Tourism: A Gender Analysis,* Chichester: John Wiley.

Eliade, M. (1957/1987) *The Sacred and the Profane,* Orlando: Harcourt Brace.

Fürst, E. L. (1995) *Mat – et annet språk. Rasjonalitet, kropp og kvinnelighet,* Oslo: Pax.

Fürst, E. L. (1998) Poststrukturalisme og fransk feminisme. Julia Kristeva i fokus, in E. L. Fürst and Ø. Nielsen (eds) *Modernitet. Refleksjoner og idébrytninger,* Oslo: Cappelen Akademisk Forlag.

Green, M. (1991) *The Sun-Gods of Ancient Europe,* London: Batsford.

Irigaray, L. (1993) *An Ethics of Sexual Difference,* London: The Athlone Press.

Jakobsen, J. K. S. (1997) The making of an attraction. The case of North Cape, *Annals of Tourism Research.*

Jasen, P. (1995) *Wild Things. Nature, Culture and Tourism in Ontario, 1790–1914,* Toronto: University of Toronto Press.

Kristeva, J. (1984) *Revolution in Poetic Language,* New York: Columbia University Press.

Kristeva, J. (1986a) Women's time, in T. Moi (ed.) *The Kristeva Reader,* Oxford: Blackwell.

Kristeva, J. (1986b) A new type of intellectual: the dissident, in T. Moi (ed.) *The Kristeva Reader,* Oxford: Blackwell.

Kristeva, J. (1986c) About Chinese women, in T. Moi (ed.) *The Kristeva Reader,* Oxford: Blackwell.

Kristeva, J. (1987) *Tales of Love,* Columbia University Press: New York.

Kristeva, J. (1991) *Strangers to Ourselves,* Harvester Wheathsheaf: New York.

Kristeva, J. (1996) *Om fallosens fremmedhet. Om hysterisk tid,* Senter for kvinneforskning, arbeidsnotat 6/96, Universitetet i Oslo.

Lauter, E. (1984) *Women as Mythmakers*, Bloomington: Indiana University Press.

Lefebvre, H. (1991) *The Production of Space*, Oxford: Blackwell.

Mills, S. (1991) *Discourses of Difference*, London: Routledge.

Myrhaug, M.-L. (1997) *I modergudinnens fotspor*, Oslo: Pax.

Pratt, M. L. (1988) *Imperial Eyes*, London: Routledge.

Reusch, H. (1895) *Folk og natur i Finnmarken, Kristiania*, A.W. Brøggers Bogtrykkeri.

Rose, G. (1993) *Feminism & Geography*, Minneapolis: Minnesota University Press.

Skavhaug, K. (1990) *The North Cape: Famous Voyages from the Time of the Vikings to 1800*, Honningsvåg: Nordkapplitteratur/Nordkappmuseet.

Smith, A. (1996) *Julia Kristeva. Readings of Exile and Estrangement*, London: Macmillan.

Steinsland, G. (1997) *Eros og død i norrøne myter*, Oslo: Universitetsforlaget.

Von Franz, M. L. (1995) *Creation Myths*, Boston: Shambala.

3

SENSING LEISURE SPACES

John Urry

Introduction

What are the most important sensuous mediators between the 'social' and the 'natural' worlds? What is the role of the senses in leading people to interpret certain places as 'natural' and others as 'unnatural'? How do the various senses enable people to distinguish between acceptable and unacceptable changes? What senses are involved in the production of different leisure spaces, such as gardens, parks, beaches, moorlands and so on? How are different leisure spaces actually sensed by visitors?

There are two useful books within which I will frame my analysis of the role of the senses and leisure. First, there is Rodaway's *Sensuous Geographies*, in which he seeks to bring together the analyses of body, sense and space (1994). He argues that all the senses are geographical or spatial. Each contributes to people's orientation in space; to an awareness of spatial relationships; and to the appreciation of the qualities of particular places. It also follows that the senses are intricately tied up with the construction and reproduction of different natures. Natures and their perception are in part produced by specific concatenations of the senses. There are five distinct ways in which the different senses are connected with each other in order to produce a sensed environment: co-operation between the senses; hierarchy; sequencing of one sense after the other; thresholds of effect of one or other sense; and reciprocal relations of some senses with the environment in question (1994: 36–7). Each of the senses has given rise to a rich array of metaphors which attest to the relative importance of each of the senses within everyday life.

The second book is Lefebvre's *The Production of Space* (1991). He argues first, that space is produced not just given, and in particular space is socially produced. So we should talk about social spatialisations. Second, he distinguishes between the domination of nature and the appropriation of nature, and hence between dominated spaces and appropriated spaces. Dominated spaces involve the destruction of nature. Appropriated spaces are those involving its 'consumption'. These are qualitatively distinct regions exploited for the purpose of and by means of consumption. Leisure spaces involving the supposedly natural phenomena of the sun, sea, snow and so on, are pertinent examples of appropriated spaces. Third, spaces exhibit an: 'increasingly pronounced visual character. They are made with the visible in mind . . . The predominance of visualization . . . serves to conceal repetitiveness. People *look*, and take sight, take seeing, for life itself' (1991: 75–6).

He describes how the visual sense has gained the upper hand and all impressions derived from the other senses gradually lose clarity and then fade away (Lefebvre 1991: 286). Social life becomes the decipherment of messages by the eye; and the eye relegates objects to the safe distance and renders them passive. The hegemonic role of visuality overwhelms the whole body and usurps its role. The embodied nature of our relationship to the world has come to be narrowly 'focused' upon the visual sense which incorporates the 'spectacularisation of life'.

Rodaway and Lefebvre can be generalised. Societies have intersected with their respective 'physical environments' in diverse ways within different historical periods. We can distinguish between four such intersections:

- the *stewardship* of particular areas of land or natural resources so as to provide a better inheritance for future generations who live within a particular local area;
- the *exploitation* of land and other resources through seeing nature as separate from society and available for its maximum instrumental destruction;
- the environment is subject to *scientisation* through treating it as the object of scientific investigation and hence to systematic intervention and regulation;
- the *consumption* of the physical environment, particularly through turning it into a garden, landscape or seascape, not primarily for production, but for appropriation.

Humans sense such environments in diverse ways. This is not a matter of individual psychology but of socially patterned 'ways of sensing'. In the following I will be concerned with sight, as in Berger's ways of seeing, but also with ways of sensing smell (1972).

The particular fascination with the eye as the apparent mirror of nature and a more general 'hegemony of vision' has characterised Western social thought and culture over the past few centuries (Rorty 1980). Such a dominance of the eye was the outcome of a number of developments across Europe (see Hibbitts 1994; Macnaghten and Urry 1998: Chap. 4 on much of the following):

- ecclesiastical architecture of the Mediaeval period which increasingly allowed large amounts of light to filter through brightly coloured windows;
- the growth of heraldry as a complex visual code denoting chivalric identification and allegiance;
- the notion of linear perspectivism in the fifteenth century in which three-dimensional space came to be represented on a two-dimensional plane;
- the development of the science of optics and the fascination with the mirror as both object and metaphor;
- the growth of an increasingly 'spectacular' legal system characterised by colourful robes and courtrooms;
- the invention of the printing press which reduced the power of the oral/aural sense.

Within Western philosophy sight has been regarded as the noblest of the senses, as the basis of modern epistemology. Arendt summarises the dominant tradition: 'from the very outset, in formal philosophy, thinking has been thought of in terms of *seeing*' (1978: 110–1).

> It is pictures rather than propositions, metaphors rather than statements, which determine most of our philosophical convictions . . . the story of the domination of the mind of the West by ocular metaphors.
> (Rorty 1980: 12–13)

Vision has also played a crucial role in the imaginative history of Western culture. Jay points out the clusters of image which surround the sun, moon, the stars, mirrors, night and day and so on; and the ways in which basic visual experience has helped to construct efforts to make sense of both the sacred and the profane (1986, 1993). Indeed such vision is by no means merely secular. Nineteenth-century myths of the American wilderness and its 'loss' and fall from innocence were shot through with religious symbolisms (see Novak 1980). Typical American landscapes were seen to involve the complex intertwinings of God and nature. Indeed nature was presumed to be something viewed through the eyes of an all-seeing God and not just through the eyes of humans or even of the landscape artists of the pre-Civil War period. Novak argues that there were various myths of nature – primordial Wilderness, Garden of the World, the original Paradise, and Paradise regained – derived from Christian theology. Most were based upon two presumptions: that nature and God were indissoluble, and that there was a realm of human practice outside and beyond such a nature that was less worthy in God's eye.

More generally, there have been complex connections between visualism and the discovery and recording of nature as something separate from human practice. Before the eighteenth century travel to other environments had been largely based upon discourse and especially on the sense of hearing and the stability of oral accounts of the world (Adler 1989). Particularly important was the way in which the *ear* had provided scientific legitimacy. But this gradually shifted as 'eyewitness' observation of nature as the physical world became important. Observation rather than *a priori* knowledge of mediaeval cosmology came to be viewed as the basis of scientific legitimacy; and this subsequently developed into the foundation of the scientific method of the West. Sense-data was typically thought of as that data produced and guaranteed by the sense of sight.

The scientific understanding of nature organised through 'scientific' travel and observation thus came to be structurally differentiated from leisured travel as such. The latter required a different discursive justification. And this was increasingly focused around, not science, but connoisseurship, of being an expert *collector* of works of art and buildings, of flora and fauna and of landscapes. In particular, travel became more obviously bound up with the comparative aesthetic evaluation of different natures, of flora and fauna, landscapes and seasides and less with eyewitness scientific observation. Travel entailed a different kind of vision and hence of a different visual ideology. Adler summarises:

> travellers were less and less expected to record and communicate their emotions in an emotionally detached, impersonal manner. Experiences of beauty and sublimity, sought through the sense of sight, were valued for their spiritual significance to the individuals who cultivated them.
>
> (1989: 22)

The amused eye increasingly turned to a variety of natures which travellers were able to compare with each other and through which they developed a discourse for what one might term the comparative connoisseurship of nature, including the knowledge of different plants. Barrell argues that such upper class leisured travellers: 'had experience of more landscapes than one, in more geographical regions than one; and even if they did not travel much, they were accustomed, by their culture, to the notion of mobility, and could easily imagine other landscapes' (1972: 63).

Gradually these other landscapes came to influence the landscapes at home, as gardens developed reflecting such other landscapes and people's ability to domesticate 'nature'. Garden designers were increasingly employed to produce new landscapes to surround the houses of the affluent – gardens to be visually consumed and thereby to convey civilised images of a now-tamed nature. As Tuan points out, such gardens were viewed as though they were landscape paintings and did not particularly emphasise or develop other senses, especially that of smell (1993: 67). Indeed, with the extensive enclosures of the land beyond the garden there was often a continuity of scenery between the garden inside and the tamed nature outside (Bermingham 1986). Much landscape painting showed the owner of the land exhibiting a visual possessive mastery over 'his' land.

This increasing hegemony of vision in European societies produced a transformation of nature as it was turned into spectacle. Paglia makes this point in dramatic fashion:

> There is, I must insist, nothing beautiful in nature. Nature is a primal power, coarse and turbulent. Beauty is our weapon against nature . . . Beauty halts and freezes the melting flux of nature.
>
> (1990: 57)

During the nineteenth century there was a general growth in viewing nature as spectacle. The nineteenth century was one of the most visual periods in Western culture; Ruskin claimed that the 'greatest thing a human soul ever does in this world is to see something . . . To see clearly is poetry, prophecy, and religion' (quoted Hibbitts 1994: 257).

Nature increasingly became seen as scenery, views, perceptual sensation and romanticised. Green suggests that partly because of the writings of the romantics: 'Nature has largely to do with leisure and pleasure – tourism, spectacular entertainment, visual refreshment' (1990: 6). The different discourses of the visual legitimated the turning of different 'natures' into spectacles. In England, in the nineteenth century, this included the awesome sublime landscapes of the Lake District, the sleek, well-rounded beautiful landscapes of the Sussex Downs, and the picturesque irregular and dilapidated southern villages full of thatched roofs, each of which generated their garden equivalent.

Green demonstrates some effects of this spectacle-isation of nature in the region surrounding mid-nineteenth century Paris. There was a prolonged invasion of surrounding regions by and for the Parisian spectator (Green 1990: 76). Two 'leisure' developments facilitated this invasion: the short trip out of the city, and the increased ownership of country houses. These combined to generate around Paris what he terms a 'metropolitan nature'. These are spaces outside the city that can be easily accessed. Such spaces are turned into safe sites for leisure and recreation for the city dweller to visit from time to time. They produce what the visitors believe is an individualised rejuvenating or refreshing experience of nature, especially through strolling in the gardens surrounding such country homes. The advertising for houses in the countryside near Paris in this period brings out the importance of the visual spectacle:

The language of views and panoramas prescribed a certain visual structure of the *nature* experience. The healthiness of the site was condensed with the actual process of looking at it, of absorbing it and moving round it with your eyes.

(Green 1990: 88)

In the twentieth century the sense of sight has been transformed by what Sontag calls the promiscuous practices of photography (1979; see Crawshaw and Urry 1996, on the following). Adam summarises:

The eye of the camera can be seen as the ultimate realisation of that vision: monocular, neutral, detached and disembodied, it views the world at a distance, fixes it with its nature, and separates observer from observed in an absolute way.

(1995: 8)

Photography has been enormously significant in democratising various kinds of human experience, making notable whatever is photographed (Barthes 1981: 34). Photography also gives shape to the very processes of travel so that one's journey consists of being taken from one 'good view' to capture on film, to another good view (Urry 1990: 137–40). It has also helped to construct a twentieth-century sense of what is appropriately aesthetic and what is not worth 'sightseeing'; excluding as much as it includes (Taylor, 1994; Parr 1995, on some recent photographic attempts to contest such a dominant aesthetic).

Photographic practices also reinforce and elaborate dominant visual gazes, especially that of the male over the landscape/bodyscape of the female. More generally photography produces the extraordinary array of circulating signs and images which constitutes the visual culture of the late twentieth century. Heidegger argues that what characterises the modern world is the 'modern world picture'; this does not mean a picture of the world, but that the 'fundamental event of the modern age is the conquest of the world as picture' (1977: 134).

Moreover, photography is associated with particular kinds of visual consumption. It is possible to distinguish five such varieties:

Romantic	Solitary
	Sustained immersion and sense of awe
	Gaze involving the sense of the auratic landscape
Collective	Communal activity
	Series of shared encounters
	Gazing at the familiar with people who are also familiar

39

Spectatorial	Communal activity
	Series of brief encounters
	Glancing and the collecting of many different signs of the environment
Environmental	Collective organisation
	Sustained and didactic
	Scanning to survey and inspect nature
Anthropological	Solitary
	Sustained immersion
	Scanning and active interpretation of the 'culture'

Finally, here we should note that sight, which is often seen as producing illumination and clarity (enlightenment), also produces its dark side. Many of the most powerful systems of modern incarceration in the twentieth century involve the complicity of sight in their routine operations of power.

In recent debates, some of the objects seen by the sightseer, including paradigmatically Disneyland, are taken as illustrating hyper-reality, forms of simulated experience which have the appearance of being more 'real' than the original (Baudrillard 1981; Eco 1986; Rodaway 1994; Bryman 1995). Such places rest upon hyper-sensuous experiences in which certain senses, especially that of vision, are seen as reduced to a limited array of features, exaggerated and then dominating the other senses. This hyper-reality is characterised by surface such that a particular sense is seduced by the most immediate and constructed aspect of the scene in question, the 'eye' at a Disneyland or the 'nose' and the sense of smell at a Fishing Heritage Centre (as at Grimsby in North-East England). This is a world of simulation rather than representation, a world where the medium has become the message.

A critique of some ocularcentric regimes has also been developed within feminist theorising. It is argued that the concentration upon the visual, or at least the non-baroque versions of the visual, over-emphasises appearance, image and surface. Irigaray argues that in Western cultures: 'the preponderance of the look over the smell, taste, touch and hearing has brought about an impoverishment of bodily relations. The moment the look dominates, the body loses its materiality' (1978: 123).

Thus the emphasis upon the visual reduces the body to surface and marginalises the sensuality of the body. In relationship to nature it impoverishes the relationship of the body to its physical environment and over-emphasises masculinist efforts to exert mastery, whether over the female body or over nature. By contrast it is claimed that a feminist consciousness less emphasises the dominant visual sense and seeks to integrate all of the senses in a more rounded consciousness not seeking to exert mastery over the 'other'. Especially significant is the sense of touch to female sexuality. Irigaray argues that: 'Woman takes pleasure more from touching than from looking, and her entry into a dominant scopic regime signifies, again, her consignment to passivity: she is to be the beautiful object of contemplation' (cited in Jay 1993: 531).

I turn now to one of these other senses, that of smell. In the nineteenth-century city, Stallybrass and White argue: 'the city . . . still continued to invade the privatised body and household of the bourgeoisie as smell. It was, primarily, the sense of smell which enraged social reformers, since smell, whilst, like touch, encoding revulsion, had a pervasive and invisible presence difficult to regulate' (1986: 139).

The upper class in nineteenth-century British cities experienced a particular 'way of sensing' such cities, in which smell played a particularly pivotal role. The romantic construction of nature was powerfully forged through the odours of death, madness and decay which, by contrast with nature, were ever-present in the industrial city (Tuan 1993: 61–2; Classen, Howes, Synnott 1994: 165–9). Corbin talks of the 'stench of the poor' in nineteenth-century Paris (1986: Chapter 9, 1992; Porteous 1985, 1990).

More generally there is a denigration of the sense of smell within Western culture – smell is thought to be more developed amongst so-called savages than amongst those who are apparently civilised (see Tuan 1993: 55–6). However, Lefebvre for one argues that the production of space is crucially bound up with smell. He says that 'where an intimacy occurs between "subject" and "object", it must surely be the world of smell and the places where they reside' (1991: 197). Olfaction seems to provide a more direct and less premeditated encounter with the environment that cannot be turned on and off. It provides a rather unmediated sense of the surrounding environment, and hence can often figure in contemporary environmental discourse. Tuan argues that the directness and immediacy of smell provides a sharp contrast with the abstractive and compositional characteristics of sight (1993).

What thus needs investigation are the diverse 'smellscapes' which organise and mobilise our feelings about particular places. The concept of smellscape effectively brings out how smells are spatially ordered and place-related (Porteous 1985: 369). In particular, the olfactory sense seems particularly important in evoking memories of very specific places; as Tuan notes, the momentary smell of seaweed was sufficient to invoke his childhood memories while sights from his childhood haunts were not (1993: 57). And even if we cannot identify the particular smell it can still be important in helping to create and sustain our sense of a particular place. It can generate both revulsion and attraction; and as such it can play a major role in constructing and sustaining major distinctions of taste. Rodaway effectively summarises what he describes as the geography of smell:

41

the perception of an odour in or across a given space, perhaps with varying intensities, which will linger for a while and then fade, and a differentiation of one smell from another and the association of odours with particular things, organisms, situations and emotions which all contribute to a sense of place and the character to places.

(1994: 68)

He summarises a variety of literary writings which have sought to characterise different places in terms of their apparent smells, including many drawn from 'nature'. G. K. Chesterton, for example, writes of 'the brilliant smell of water, the brave smell of stone, the smell of dew and thunder, the old bones buried under' (quoted Rodaway 1994: 73). Tolstoy describes the smells following a spring thunderstorm: 'the odour of the birches, of the violets, the rotting leaves, the mushrooms, and the wild cherry' (quoted Tuan 1993: 62).

Has the sense of smell become less significant in modern Western societies, as Lefebvre appears to argue (1991)? In many pre-modern societies the sense of smell has clearly been very significant (Classen, Howes, Synnott 1994). In modern societies there is an apparent dislike of strong odours. Public health systems separate water from sewerage; a lack of smell indicates personal and public cleanliness, very frequent baths and showers are favoured; perfume is restricted to adult women. To be sensitised to different smells and to have some ability to prevent those smells deemed to be 'unnatural' (which may of course include many smells such as rotting vegetables which are 'natural') is taken to be important.

More generally, Bauman argues that modernity: 'declared war on smells. Scents had no room in the shiny temple of perfect order modernity set out to erect' (1993: 24). Modernity sought to neutralise smells by creating zones of control in which the senses would not be offended. Zoning became an element of public policy in which planners accepted that repugnant smells are in fact an inevitable by-product of urban-industrial society. Thus refuse dumps, sewage plants, meat-processing factories, industrial plants and so on are all spaces in which bad smells are concentrated, and are typically screened-off from everyday life by being situated on the periphery of cities. This notion of the war on smells in modernity was of course carried to the extreme in the Nazi period, where the Jews were routinely referred to as 'stinking' and their supposed smell was associated with physical and moral corruption (Classen, Howes, Synnott 1994: 170–5).

But Bauman argues that smell is a particularly subversive sense since it cannot be wholly banished (1993). It reveals the artificiality of modernity. The modern project to create a pure, rational order of things is undermined by the sweet smell of decomposition which continuously escapes control and regulation. Thus the 'stench of Auschwitz' could not be eliminated even when at the end of the war the Nazis tried to conceal what had happened (Classen, Howes, Synnott

1994: 175). This is why Bauman submits that decomposition has 'a sweet smell'. As one of the liberators at Auschwitz wrote:

> The ovens,
> the stench,
> I couldn't repeat
> the stench. You
> have to breathe.
> You can wipe out
> what you don't want
> to see. Close your
> eyes. You don't want
> to hear, don't want
> to taste. You can
> block out all the senses
> except smell
>
> (cited in Classen, Howes, Synnott 1994: 175)

But some of the attempts to restrict and regulate smell have gone into reverse in the past couple of decades in the 'West', as the cultural turn to 'nature' has become pronounced. Recent leisure-related trends include (see Macnaghten and Urry 1998: Chap. 4, for more detail):

- the increased attraction of spicy 'oriental' foods (a matter of smell and taste);
- the increased use of natural, often eastern, perfumes both for the body for both genders and for the home;
- a reduced emphasis upon antiseptic cleanliness and a greater use of materials and smells that are deemed to be 'natural' (such as lemon);
- much greater knowledge of and sensitivity to the smells of nature, especially flowers;
- a greater awareness of the malodorous smells of the motor car and of the many chemicals identifiable in rivers and seas which may be taken as metonymic of much wider and long-term risks;
- some appreciation that the preservation of landscapes involves not just issues of visualisation, but also of threatened smellscapes.

The cultural shift to nature has increased the power of smell over and against that of sight; but this does not mean that all smells are presumed to be equally acceptable (cigarette smoke, nitrates and sewerage are not). The cultural shift to the natural thus ushers in new contestations over diverse smellscapes as they interact in complex ways with the dominant processes of visualisation.

I have thus touched upon a number of issues concerned with the complex, ambivalent and contested processes by which we encounter various 'natures' and 'spaces'. It is my claim here that leisure spaces are to be analysed in terms of the categories examined here, categories which hopefully provide a more sophisticated discourse by which people's pleasures in leisure can be interrogated and even enhanced.

References

Adam, B. 1995a. 'Radiated identities: in pursuit of the temporal complexity of conceptual cultural practices', *Theory, Culture and Society Conference*, Berlin, August.

Adler J. 1989. 'Origins of sightseeing', *Annals of Tourism Research*, 16: 7–29.

Arendt, H. 1978. *The Life of the Mind*. New York: Harcourt Brace Jovanovich.

Barrell, J. 1972. *The Idea of Landscape and the Sense of Place. 1730–1840*. Cambridge: Cambridge University Press.

Barthes, R. 1981. *Camera Lucida*. New York: Hill and Wang.

Baudrillard, J. 1981. *For a Critique of the Economy of the Sign*. St Louis: Telos.

Bauman, Z. 1993. *Postmodern Ethics*. London: Routledge.

Berger, J. 1972. *Ways of Seeing*. Harmondsworth: Penguin.

Bermingham, A. 1986. *Landscape and Ideology*. London: Thames and Hudson.

Bryman, A. 1995. *Disney and his Worlds*. London: Routledge.

Classen, C., Howes, D. and Synnott, A. 1994. *Aroma: The Cultural History of Smell*. London: Routledge.

Corbin, A. 1986. *The Foul and the Fragrant*. Leamington Spa: Berg.

Corbin, A. 1992. *The Lure of the Sea*. Cambridge: Polity.

Crawshaw, C. and Urry, J. 1996. 'Tourism and the photographic eye', in C. Rojek and J. Urry (eds) *Touring Cultures*. London: Routledge.

Eco, U. 1986. *Travels in Hyper-Reality*. London: Picador.

Green, N. 1990. *The Spectacle of Nature*. Manchester: Manchester University Press.

Heidegger, M. 1977. *The Question Concerning Technology and Other Essays*. New York: Harper Torchbooks.

Hibbitts, B. 1994. 'Making sense of metaphors: visuality, aurality, and the reconfiguration of American legal discourse', *Cardozo Law Review*, 16: 229–356.

Irigaray, L. 1978. 'Interview with L. Irigaray', in M.-F. Hans and G. Lapouge (eds) *Les femmes, la pornographie et l'érotisme*. Paris: Minuit.

Jay, M. 1986. 'In the empire of the gaze: Foucault and the denigration of vision in twentieth-century French thought', in D. Hoy (ed.) *Foucault: A Critical Reader*. Oxford: Blackwell.

Jay, M. 1993. *Downcast Eyes*. Berkeley: University of California Press.

Lefebvre, H. 1991. *The Production of Space*. Oxford: Blackwell.

Macnaghten, P. and Urry, J. 1998. *Contested Natures*. London: Sage.

Novak, B. 1980. *Nature and Culture: American Landscape and Painting, 1825–1875*. London: Thames and Hudson.

Paglia, C. 1990. *Sexual Personae*. Harmondsworth: Penguin.

Parr, M. 1995. *Small World*. Stockport: Dewi Lewis.

Porteous, J. 1985. 'Smellscape', *Progress in Human Geography*, 9: 356–78.

Porteous, J. 1990. *Landscapes of the Mind: Worlds of Sense and Metaphor*. Toronto: Toronto University Press.

Rodaway, P. 1994. *Sensuous Geographies*. London: Routledge.

Rorty, R. 1980. *Philosophy and the Mirror of Nature*. Oxford: Blackwell.

Sontag, S. 1979. *On Photography*. Harmondsworth: Penguin.

Stallybrass, P. and White, A. 1986. *The Politics and Poetics of Transgression*. London: Methuen.

Taylor, J. 1994. *A Dream of England*. Manchester: Manchester University Press.

Tuan, Y-F. 1993. *Passing Strange and Wonderful*. Washington, D.C.: Island Press.

Urry, J. 1990. *The Tourist Gaze*. London: Sage.

4

PARKS AND GARDENS: METAPHORS FOR THE MODERN PLACES OF SPORT

John Bale

Introduction

Achievement-oriented sport is widely regarded as a product of the modern age or as a paradigm of modernity. It is hardly surprising, therefore, that metaphors used to describe sports are often borrowed from the world of science and technology. For example, some neo-Marxist observers have likened sports tactics to 'spatio-temporal planning models' (Rigauer 1993: 289) and the game of football to a 'machine' (Vinnai 1973). Likewise, stadia could be metaphorically described as 'assembly lines' for the 'production' of record outputs. Sports places are indeed a world of mathematics and science, precisely measured segments and territories which seek to neutralise the vagaries of nature (Bale 1993). Other metaphors for the modern sports place include 'container' or even 'prison'. Indeed, Brohm regards sport as perhaps the social practice which best illustrates the disciplinary society of Michel Foucault (Brohm 1978). After all, is the stadium not an 'enclosed, segmented space, observed at every point, in which the individuals are inserted in a fixed place, in which the slightest movements are supervised, in which all events are recorded' (Foucault 1979: 197)? Adopting as they do the 'hard', masculinised metaphors from the world of science, such readings of the stadium and other sports places present an unambiguous view of sport as anti-nature, as places of dominance.

In common with other recent observers of the ways in which the landscape may be represented, I want in this paper to use a 'softer', feminised metaphor borrowed from the humanities (Cosgrove and Domosh 1993: 30). As an alternative reading I will refer to sports landscapes as gardens, not in order to hark back to some mythical idyll, but to emphasise the *ambiguity* of sports landscapes and the way in which they reflect the exercise of both dominance *and* affection. This implies, of course, that the garden itself is an ambiguous and anomalous category between nature and culture. Parks and gardens are not precisely the same, but each connotes certain similar qualities. A park implies a broader spatial extent than a garden, the latter tending to be enclosed within the former. But both are 'improvements' on nature and both connote – among other things – leisure and playfulness.

Of the ninety-two football clubs that make up the major leagues in the English game, twenty-five play their home games in places denoted as 'parks'. Among the most well known are Villa Park (Birmingham) and Goodison Park (Liverpool). In North America two famous sports arenas are named Madison Square Garden (New York) and Maple Leaf Gardens in Toronto; throughout the same continent there are a large number of sports facilities called 'field houses'. But the visitor will see no sign of any trees in Villa Park, no shrubs in Madison Square Garden and no meadows in the field houses. And those who *work* at the highly rationalised activities which take place in these parks and gardens are *players*. Sports *parks* and *gardens* are therefore spaces whose instability and ambiguity are clearly evident through the words themselves. Sports parks are often made of steel, concrete and plastic; defined by geometric shapes, straight lines and right angles. Although parks and gardens are not exactly the same, they do both invoke a pastoral milieux and carry benign connotations (Samuels 1979). In other words, a single or conventional *meaning* is not immediately present in the signs 'park' or 'garden'.

There is clearly no way that such signs, when being used to describe a modern sports stadium, can be said to be passing themselves off as accurate descriptions of what we see. It may appear that they are either residual terms which (in Britain and North America, at least) have resisted replacement by the more powerful (and equally value-laden term connoting image upgrading and power) word 'stadium'. They may also be ideologically motivated terms encouraging the illusion of an idyllic past. Certainly, when reading the iconographic representations of the landscape of English cricket (in painting and prose) this latter reading might be encouraged (Bale 1993: 163–5), but I will not follow these lines of argument in the present paper. Instead, I will suggest that sports places, including the most anonymous of football stadia, possess many of the characteristics of the more conventional 'parks' and 'gardens' and that the sign 'garden' *is not* so distant from its more conventional, etymological origin as might at first be thought. A reading of the modern stadium *as* a park or garden reveals not only its ambiguous character; it also provides a softer image than that projected by the use of alternative metaphors.

My basic framework for this paper is related to the work of the North American cultural geographer, Yi-Fu Tuan. Tuan's work has consistently run counter to the positivist trend in post-war human geography and he has promoted a more phenomenological view of place and landscape by means of his 'descriptive psychological geography' (Tuan 1984). In one of his more pessimistic essays, *Dominance and Affection: The Making of Pets*, Tuan approaches the landscapes of gardens and pets as reflections of the exercise of human power over nature in botanic and zoological contexts (Tuan 1984). Such an approach could also be applied to sports places. First, however, I want to briefly review the notion that sport, and hence the places where sports are practised, is anti-nature. I will then move on to an application of the garden metaphor, taking in several examples of sports landscapes which illustrate the ambivalent character of the sport-garden.

Sport as anti-nature

The notion of the sports place as a machine or prison makes it appear anything but a garden; it certainly implies that it is opposed to nature. But sport, as part of civilisation itself, 'could only emerge through the mastery of nature' (Tuan 1984: 173). The way in which sports landscapes have become increasingly enclosed, segmented and artificial could be said to reflect the way in which the links between nature and everyday life have been battered by modernity [so that] for the most part, nature and society are now perceived to be counterpoised (Katz and Kirby 1991).

During the eighteenth and nineteenth centuries the landscape of sport became increasingly artificial as science and technology 'improved' the cultivation of sports, just as they were improving the cultivation of wheat, barley, sheep and cattle. Such improvement resulted in the emergence of *sportscape* – a landscape *given over solely to sport*, and a manifestation of modern sport's fixation with neutralising or altering the effects of the physical environment. Sportscape exemplifies a fixation with improving on nature and artificialising the landscape in its quest for the optimal sporting environment within which performances in different places could be meaningfully compared. Such improvement, essential for record purposes, logically dictates that the world of sport ought to be one of 'placelessness', one sports place being the same – exactly the same – as any other.

Early competitive sports, even in their modern, codified forms, were played, literally, in fields, parks and gardens. English cricket was played in farmers' fields; football took place in municipal parks; tennis was played in middle class domestic gardens. But the notion of sport as a form of athletic 'production' sets it against nature and during the late nineteenth century monofunctional spaces emerged in order to satisfy the dictates and requirements of modern sports. In a discussion of the problems of sport, ecology and progress, it is therefore not surprising to read how the human body and nature are each scientifically modified in order to fulfill the ideology of sport. The:

'correcting' of nature is justified according to requirements of sport, confirmed by the general ideology of growth and progress. Whenever our body or nature is not given the biologically necessary periods of maturation and recreation this is artificially affected. In the eyes of sports bureaucrats, a further 'improvement' of nature is necessary when the basic conditions for training and competitions have to include equal chance for all competitors. Equal chance and tensions as constituting elements of sport require constant conditions throughout a certain period . . . Nature with its processes of growth, its variability and its biologically determined regeneration periods, is not equipped to comply with these conditions . . . These elements of uncertainty of man [sic] or nature have to be extinguished to attain the scientifically calculated goal.

(Peyker 1993: 73)

It has also been noted that in sport

real play and enjoyment, contact with air and water, improvisation and spontaneity all disappear. These values are lost to the pursuit of efficiency, records, and strict rules. Training in sports makes of the individual an efficient piece of apparatus which is henceforth unacquainted with anything but the harsh joy of exploiting his [sic] body.

(Ellul 1965: 383)

These unambiguous stances reflect the way in which the landscapes of sports are dehumanised (or over-humanised) and anti-nature, going far beyond simply killing plants or animals for survival. It can be argued that sport (though not disport) is one of the many things that are done once survival is assured and is a plausible reality only to people who have satisfied their more pressing needs. (Tuan 1984: 18) The formal landscaping found in both gardens and sports places reveals the human need to dominate (ibid.: 22). To allow for a humanised world we 'use power to change nature for our own benefit and delight' (Tuan 1986: 128), i.e. to break a tree to make a spear in order to kill an animal for survival – or to cut grass in order for a ball to roll smoothly as in football. As Cosgrove notes, 'humanism has always been very closely aligned in practice with the exercise of power both over other human beings and over the natural world' (Cosgrove 1989: 119). The problem seems to be that although a humanised world may be inevitable, an over-humanised world need not be. A humanised world can give great pleasure but an excess of humanism can paradoxically produce a dehumanised landscape (Relph 1981). The metaphor of the garden can therefore be used to illustrate the ambiguity, not only of the sports place, but of power itself.

49

In sport, as in play, 'the abuse of power is evident less in any quantitative measure as in the character of change – in the ways that power has been used to distort plant, animal and human nature to aesthetic [and sporting] ends, and the ways that animals and humans . . . [as athletes] . . . have been used to suffer indignities and humiliation' (Tuan 1984: 4). But, there is considerable variation in the extent of such distortion of nature and it is to examples of such variation that I now want to turn.

Sportscapes as parks and gardens

Tuan's ideas are undoubtedly appropriate for a more ambiguous reading of the sports landscape than that presented above. Whereas, at a certain level of simplicity, the domination of nature in the production of arable and pastoral landscapes results from the basic need for human survival, the football field, golf course, sports hall, ski jump, tennis court and swimming pool – like ornamental gardens and artificial lakes – are landscapes which have been transformed in order to provide pleasure once the necessity for food has been secured. They are, therefore, forms of power or domination, but they do not necessarily result from a hatred of nature but, rather, a desire to civilise it. They may, after all, still be like the parks and gardens out of which their modern forms grew, exemplifying what Tuan calls 'playful domination' (ibid.: ix). And such sports places are certainly treated with affection. Witness the love that sports fans have for their favourite stadia and arenas, and the resistance they apply when such places are threatened with 'development' (Bale 1994: op. cit). The love of the youthful, male English football fan for *his* stadia is no less irrational than the love of an ageing, homebound lady for *her* window garden. Indeed, sports places, like gardens, are often viewed as 'cultural models of the good life' (Tuan 1986 op. cit: 25). This illustrates the ambivalent nature of power and the relationship between dominance, on the one hand, and affection on the other.

Throughout *Dominance and Affection* Tuan stresses that dominance is not the opposite of affection but is, in fact, dominance with a human face. Dominance may be cruel and exploitative, with no hint of affection in it. What is produced is the victim. This would summarise the deep ecological view which sees nature as an exploited victim of the rationalised, artificial, isotropic plane of sportscape, typified by the austere, concrete stadium or the synthetic running track. But dominance may be combined with affection, and what it then produces is the garden (Tuan 1984: 2). This reflects a more benign, but ambiguous, use of human power. As Tuan notes, 'affection mitigates domination, making it softer and more acceptable, but affection itself is only possible in relationships of inequality' (Tuan 1986: 5).

It is this softer form of power (that is, affection), often (though not always) reflected in the landscape and parks of achievement sport, that I will use to illustrate these ideas. My examples range from sports landscapes which, while illustrating the exercise of power, have minimal impacts on the natural landscape, to those where the power relation is one of true dominance. Between these two extremes, the sports landscape is, like the modern garden, one of 'playful domination'.

Taming the field of play; the stadium and other sportscapes.

Despite sports participants' apparently benign use of the natural landscape in such sports as orienteering or canoeing, they do, nevertheless, *use* it (Tuan 1984: 176). Even orienteering can temporarily damage vegetation, though most of it recovers quickly. But landscapes designed specifically with sport in mind, like 'formal landscaping reveals the human need to dominate' and the domination and possession of nature for sport, as with other land uses, occurs in three stages (ibid.: 2). Prospective sportscape is first *mapped*, the landscape being summarised cartographically for purposes of marketing it to the world of various sports businesses; second, it is *mythologised* as being ideal or 'natural', as, for example, with golf courses, ski-pistes, and indeed, the continued use of the words 'garden' and 'park'. Third, it is *aestheticised* by landscape gardening, the use of exotic plants or novel architectural forms (Katz and Kirby 1991: 265). In many cases, however, aestheticisation is unconvincing and landscape and nature become dominated by sport to such an extent that it causes considerable controversy. For the 1992 Winter Olympics at Albertville in France ski runs were carved out of mountainsides leaving permanent scars. Rare high mountain marshland was damaged by cross-country skiing pistes, and an environment-friendly ski-jump was rejected and replaced by a controversial facility driven into the mountainside with 300 giant piles. Such examples illustrate the clear and brutal dominance of sport over nature. These are more analogous to concrete car parks than the 'natural' parks of the cross-country skier.

Dominance in a more benign form may be illustrated by the stadium and other sportscapes which share almost every basic characteristic of the garden or park. The green enclosures of sports, like many gardens, may be reminiscent of university courtyards and quadrangles, but seem to me more redolent 'of the garden world, the condition of leisure where all play aspires to paradise . . . a green expanse, complete, coherent, shimmering, carefully tended, a garden' (Giamatti 1982). Stadia, ski slopes, tennis courts and golf courses, for example, can still be classic blendings 'of nature and artifice: [like gardens] they are the product of horticulture and architecture. On the one hand, they consist of growing things; on the other of walls, terraces, statues . . . ' (Tuan 1986: 21). Even so, in constructing such often loved facilities human beings display power to spare.

Although landscape gardeners – and their suburban hobbyist equivalents – have been deadly serious about the scientific improvement on nature, the garden has, to a large extent, always been associated with playfulness (ibid.: 29). But 'the garden is not only a product of play but also an arena *for* play' (Tuan 1989) and, as a result, in one or two cases modern sports actually grew out of playful activities traditionally associated with the garden. Tennis and croquet come most readily to mind; but in such cases the activity was hardly playful enough to leave nature unaffected since a flat surface and a meticulously maintained lawn were required. Developed from the indoor 'real' or 'royal' tennis, the burgeoning suburban gardens of early twentieth-century England formed the focal points for the growth of not only individual grass courts on the garden lawns of the suburbs, but also suburban tennis clubs with eight or more courts, sheltered from both summer sun and the intrusive gazes of the lower classes by a luxuriant foliage of trees.

The cricket field is a paradigm of the 'improvement' of nature in the name of sport. Traditionally, the preparation of the cricket pitch was managed by sheep but an improved playing surface was a 'necessary condition for the development of cricket in its modern form' (Allison 1980). In 1864 the first grounds*man* was appointed at Lord's Cricket Ground in London. The stadium and the garden became homologous as a result of the expertise of those who tended them; groundsmen actually *were* gardeners. Those of late nineteenth-century England, those whose job it was to perfect the surfaces upon which sports were 'played', were fully familiar with horticulture, often bringing their expertise from the environments of country houses to those of urban stadia. Consider, for example, the case of Percy Peake who, in 1874, went to Lord's Cricket Ground and commenced the task of making it famous the world over for the quality of its playing surface. Peake had been a gardener and a keeper of lawns before he first made a name for himself as a cricket groundsman at Brighton. He was an expert in agriculture, horticulture and geology, although he never pursued formal training in those disciplines. His careful and empirical study of soils, grasses and marls gradually produced a surface at Lord's that became the envy of every county club. He wrote many articles on the preparation of pitches and spent his last years laying down several cricket grounds across the country (Sandiford 1964).

The introduction of the heavy roller (first used at Lord's in 1870) further served to standardise the playing surface. The late nineteenth century saw the growth in production of specialised cricket equipment to aid the gradual levelling and equalising of the cricket landscape in an age of rapidly changing technology. By the first decade of the present century the groundsman had technologised the cricket pitch and was arguably *the* vital actor in the emergence of the sports landscape. Like the garden, the sportscape had to be maintained thoughtfully and systematically; otherwise it would revert to nature (Tuan 1984: 19).

The cricket and football fields exemplify the imposition of human power on a natural landscape. The stadium is the outcome of the application of horticulture and architecture, which together produce the sporting equivalent – in more austere form – of the garden-quadrangle or landscaped park. I have already alluded to some of the applications of horticulture to sportscape, but the broader landscape is completed by the work of architects whose structures – the stadia – enclose the fields of achievement upon which the sporting action takes place. An arena or stadium 'may look more like its ancient precursor than anything else in the modern world looks like its architectural ancestors' (Giamatti 1982: 32–3) and the stadium, like the garden 'is power in confident mood – witness the size and masterful layouts of some of its grander specimens' (Tuan 1984: 32). It is difficult to avoid being impressed by the power symbolised by modern Olympic stadia. That at Berlin, built for the 1936 Games, was more than a stadium and part of an integrated suburban *Reichssportfeld* which has been described as the first Olympic park complex; it 'also served to glorify the Fascist Germany and feigned a peaceful Olympic state. Its architectonic form was determined by the official state architecture' (Wimmer 1976: 35). The conglomeration of sports facilities put nature well and truly in its place. In addition to the main stadium there was an open air swimming stadium, polo field, hockey stadium, field for equestrian events, and an open air stage surrounded by a large Roman-style open air theatre where gymnastic events were held.

At the same time, however, the architecture of Olympism need not be identified with brutalism. The stadium (again like the garden) can also display 'power in a whimsical mood – witness the details of design' (Tuan 1984: 32). The Stockholm Olympic Stadium is a case in point. Designed by Torben Grut for the 1912 Olympics, it was built in the vernacular style with grey-violet brick walls and granite towers. Pierre Baron de Coubertin regarded the stadium highly and 'with its pointed arches and its turrets, its technical perfection, its good order and its purposeful disposition seemed to be a model of its kind' (Wimmer 1976: 190). It is remarkably human in scale and is today a much-loved element of the Stockholm townscape, 'a stadium part castle, part mansion, part cloister, part pageant but altogether magnificent and built entirely out of Swedish materials' (Inglis 1990). It is impossible to chronicle the stadia large and small which have architectural merit and charm or, in the case of small stadia in country towns, possess an air of innocence (ibid.: 233), often seemingly at one with the landscape out of which they have grown. Certainly, such stadia possess a sense of place or *genius loci*. My point here, however, is that while being the undeniable result of human power, sometimes evident in a boastful way, such buildings can nevertheless display whimsy and playfulness in design.

The ambitious garden-architect is willing to go to considerable lengths to create magical worlds which constantly surprise the visitor (Tuan 1984: 35). To what extent can the same be said of those who build stadia? Although many possess individual differences and some do provide extremely pleasant surprises, (Bale 1993: 78) a sports stadium has to be more than a stadium to make it magical. Stadia with surprises in them are those which are, in themselves, often ambiguous structures. The football stadium in Monaco is one (Inglis 1990); the SkyDome in Toronto is another. In this latter case nature is totally neutralised by artifice, its synthetic 'field' upon which no grass grows and its retractable roof making even the normal dualism between indoor and outdoor ambiguous. But its external mural decorations are at least whimsical while the ability of the Dome to host all manner of events, from football to funfairs makes it magical for at least some people. Like many parks, the Dome is related less to sport than to leisure activities *per se*.

All too often, however, the stadium appears as a bland, containerised structure, one looking more or less the same as any other. In the United States the term 'concrete saucer' is often applied to anonymous bowls which all look the same. In recent decades, in the context of baseball, it has been suggested that in the belief that form should follow function, builders have erected huge concrete circles and called them baseball parks. Everywhere the emphasis has been on symmetry, uniformity and keeping everything in balance (Bale 1983: 78). Such features dominate nature and often have massive car parking provisions, further contributing to the dominance of sport over the natural landscape. Where synthetic fields exist the sameness of places is even more in evidence. It is arguably the synthetic running track that has the most human pressure placed upon it to be identical to all others of its type.

In the previous sentences I have implied that in the case of many sportscapes one place increasingly – and often necessarily – becomes much the same as any other. Such landscapes produce what Edward Relph termed 'placelessness', in its purest form uniformity, 'an environment without significant places' (Relph 1976). The concrete bowl stadium with its plastic carpet and the fenced-in tennis court with its synthetic surface typify such places. In such cases, the attempt to create simple modifications of nature which will accommodate human sporting activities have, like so many gardens, failed to provide that simplicity. Gardens so often become high art and stadia often become high tech – each 'dependent upon social and technological power' (Tuan 1986: 115). Simplicity and freedom become in danger of giving way to blandness and constraint.

But it must be said that sports places vary in their degree of placelessness; some sports are closer to landscape uniformity than others. In many cases the pre-existing landscape is not simply flattened, but is sculptured to produce aesthetically pleasing slopes and hills, lakes and woods, like gardened landscapes. Flowers and shrubs, lakes and bushes, are placed around the edge of the playing area and on occasions are integral to it. Indeed, many people would regard them as more visually enjoyable than the landscapes, natural or agricultural, which preceded them.

In such cases the sports landscape becomes symbolically less *gendered* than the placeless football field. The garden ensemble may be recognised as being made up of two basic elements. The carefully homogenised grass area, traditionally tended by (grounds)men, and the borders filled with a diversity of decorative flowers. The scientifically tended lawn and the artistically composed border may be symbolically viewed as a highly gendered demarcation of garden space. Indeed, it has traditionally been the man who scientifically tended the lawn and the woman who artistically looked after the flowers (Tuan 1986). As noted earlier, the cricket pitch is a scientifically tended area of immaculately maintained lawn by the grounds*man*. On this grassed area a rational, scientific, quantitative and minutely recorded ritual is played out with utmost seriousness and concentration. Surrounding this playing (working) are(n)a, however, is often an aesthetic ensemble – to enjoy. Such sportscapes are classically exemplified by golf courses.

Golf courses provide an excellent example of the conversion of natural landscapes to sportscape through the imposition of human power over natural or semi-natural environments. Golf can, in fact, now be played across such a wide range of landscapes that it is difficult to think of the sport as possessing any general landscape characteristics – apart, of course, from holes, greens, fairways and rough. There are over 13,000 conventional golf courses in the USA, in total covering over 1,500,000 acres of land, twice the area of the state of Rhode Island (Adams and Rooney 1985). The landscape itself becomes a plaything, like landscaped parkland, often 'ingenious and full of surprises' (Tuan 1984: 34) with the designer being 'a specialist in optical illusions, in the deceptive organization of distances, in the sly concealment of obstacles' (Clay 1987). The manicured green of the superior golf course is sportscape in perfected form – 'the ultimate outdoor artifact – the prototypical American lawn', (ibid.: 219) or 'a state of nature apt to the age: a vast acreage of greenery scrupulously regulated to support a network of tiny, shallow holes' (Sorkin 1992). With the application of vast amounts of herbicides and pesticides, golf courses are gardens writ large. Although such landscapes obviously reflect the dominance of human power over nature, the dominance is ambivalent, as evidenced by the equally obvious affection with which many such landscapes are held.

Water as well as land is subjected to power relations in the name of both gardening and sport. Tuan notes how in the garden world water is manipulated for pleasure in the construction of lakes and fountains (Tuan 1984: 38). In the world of sports swimming, baths control water in typically modern form. As Charles Sprawson (quoting Jean Didion) notes, in his astonishing book *Haunts of the Black Masseur*, a swimming 'pool is, for many of us in the West, a symbol not of affluence but of order, of control over the uncontrollable. A pool is water, made available and useful' (1993). Fresh water is not required to fill the pool at regular intervals and, like the garden fountain, it is controlled to circulate and be cleansed without needing to be changed for many months. The rigid geometry of its rectangular shape reflects the essential opposition to nature and serves to encourage achievement-oriented swimming (up and down a regulation length pool in straight lines) rather than more frolicsome water movements which invariably get in the way of serious, 'sportised' swimming. In recent years, however, more play-oriented water environments have been developed in which the garden has been simulated by the presence of waterfalls, fountains and plastic trees. These playful places might be more appropriately termed water gardens than swimming pools.

Parks for play?

Gardens are widely regarded as sources of freedom and play, havens for pleasure away from a world of control and constraint. But surely play has been removed from modern sport and replaced by the deadly seriousness of the working professional? And is not the space of sport now neatly demarcated between those experts who work at playing and those *amateurs* – the spectators – who are mere observers of the spectacle from their numbered seats? To what extent does the playful character to the park or garden spill over into the modern sports park?

Although the sports park is highly rational and quantified, often synthetic and enclosed with players segregated by visible boundaries from the spectators, it cannot be said that such places signify an unambiguous form of modernity. That is to say, a sports park does not yet possess a clear differentiation in space between different types of participants (players, spectators). We do not (yet?) have the 'image of a stadium full of spectators silently watching the performance and not taking part in the drama, who consequently cannot change the results' (Archetti 1992). Indeed, there is something paradoxical about having spectators at *modern* sports events, simply because they do influence the result, hence contravening the ideal of 'fair play'. The *liminal* character of the boundaries between players and spectators means that each can play with the other; it reflects a more 'open' environment, reminiscent of the park. On the field, 'clowns' and 'fools' perform their antics – within the confines of the scientific bases of achievement sports. The fans can also fool around – shouting, singing, gesturing. Liminality is important in sports, as elsewhere, because it 'represents a liberation from the regimes of normative practice and performance codes' (Shields 1991: 84). In England, all-seat parks/stadia are as unpopular to football fans as are signs in municipal gardens saying 'Keep off the grass' to children. In other words, the strictly ordered world of rigidly defined geometrical and ordered cells which sports parks *ought to be* according to the spatial rules and regulations, turn out to be shifting inter-stitial spaces with playful dialogue between superficially segmented participants. This is not a world of the prison; it seems, after all, nearer that of the park.

Conclusion

There is no meta-metaphor which summarises accurately the landscape of modern sport. But whereas 'parks' and 'gardens' originally denoted a place for play, the words are now used in a metaphorical sense to describe a place where profes-sionals 'play' at their job. But perhaps this continued use of 'park' and 'garden' should not surprise us, given the similarities between stadia (and other sport places) and parks as outlined above. But it is Tuan's exploration of the garden (and the pet – whose sporting analogue is the athlete) as an example of the *ambiguous* character of power which is helpful in providing an alternative reading of sportscape to that dominated by 'hard' metaphors. The garden metaphor allows us to see sport as not entirely machine-like; it permits the play-element to remain in focus and does not eliminate nature and affection from a world often regarded as being dominated by science, technology and exploitation.

References

Adams R. and Rooney J. (1985) 'Evolution of American golf facilities', *The Geographical Review*, 75 (4), 419–38.

Allison L. (1980) 'Batsman and bowler: the key relation of Victorian England', *Journal of Sports History*, 7 (2), 5–20.

Archetti E. (1992) 'Argentinian football: a ritual of violence', *International Journal of the History of Sport*, 9 (2), 209–35.

Bale J. (1994) *Landscapes of Modern Sport,* London, Leicester University Press.

Bale J. (1993) *Sport, Space and the City*, London, Routledge.

Brohm J-M. (1978) *Sport: A Prison of Measured Time*, London, Ink Links.

Clay G. (1987) *Right Before Your Eyes: Penetrating the Urban Environment*, Washington, D.C., Planners Press.

Cosgrove D. (1989) 'Geography is everywhere: culture and symbolism in human geo-graphy', in D. Gregory and R. Walford (eds), *Horizons in Human Geography,* London, Macmillan, p. 119.

Cosgrove D. and Domosh M. (1993) 'Author and authority: writing the new cultural geography', in J. Duncan and D. Ley (eds), *Place/Culture/Representation*, London, Routledge, p. 30.

Ellul J. (1965) *The Technological Society*, London, Macmillan.

Foucault M. (1979) *Discipline and Punish*, Harmondsworth, Penguin.

Giamatti B. (1982) *Take Time for Paradise: Americans and their Games,* New York, Summit Books.

Inglis S. (1990) *The Football Grounds of Europe*, London, Collins.

Katz C. and Kirby A. (1991) 'In the nature of things: the environment and everyday life', *Transactions of the Institute of British Geographers,* 16 (3), 259–71.

Peyker I. (1993) 'Sport and ecology', in S. Riiskjaer (ed.), *Sport and Space,* Council of Europe, Copenhagen.

Relph E. (1976) *Place and Placelessness,* London, Pion.

Relph E. (1981) *Rational Landscapes and Humanistic Geography,* London, Croom Helm.

Rigauer B. (1993) 'Sport and the economy', in E. Dunning, J. Maguire and R. Pearton (eds), *The Sport Process,* Human Kinetics, Champaign, Ill.

Samuels M. (1979) 'The biography of landscape', in D. Meinig (ed.), *The Interpretation of Ordinary Landscapes,* New York, Oxford University Press, pp. 51–88.

Sandiford K. (1964) 'Victorian cricket technique and industrial technology', *The British Journal of Sports History,* 1 (3), 272–85.

Shields R. (1991) *Places on the Margin: Alternative Geographies of Modernity,* London, Routledge.

Sorkin M. (1992) 'See you in Disneyland', in M. Sorkin (ed.), *Variations on a Theme Park,* New York, Noonday Press, pp. 205–32.

Sprawson C. (1993) *Haunts of the Black Masseur: The Swimmer as Hero,* London, Verso.

Tuan Y-F. (1984) *Dominance and Affection: The Making of Pets,* New Haven, Yale University Press.

Tuan Y-F. (1986) *The Good Life,* Madison, University of Wisconsin Press.

Tuan Y-F. (1989) *Modernity and Imagination: Paradoxes of Progress,* Madison, University of Wisconsin Press.

Vinnai G. (1973) *Football Mania,* London, Ocean Books.

Wimmer M. (1976) *Olympic Buildings,* Leipzig, Edition Leipzig, p. 35.

5

HERITAGE AND NATIONALISM: GENDER AND THE PERFORMANCE OF POWER

Cara Aitchison

Introduction

Heritage has now become a major preoccupation of leisure and tourism studies. There has been a lack of acknowledgement, however, of the role of gender in both the representation and consumption of heritage places and products. Whilst geographical discourses have engaged with the concept of 'gendered space' in the creation of place and landscape, leisure and tourism theory is only beginning to debate issues of gender and spatialisation (Aitchison and Jordan 1998; Watson and Scraton 1998).

This chapter attempts to bridge the divide between contemporary cultural geography and leisure and tourism studies by offering a critique of gendered representations of heritage and their role in the creation of gendered spaces and places within cultural tourism. The chapter seeks to contextualise and engender heritage by teasing out the multiplicity of interconnections between space, place and gender experienced in relation to the social construction of heritage landscapes and iconographies. Instead of viewing heritage as merely occupying material space, this chapter contextualises the symbolic nature of heritage landscapes and the significance of constructing 'the gaze' and 'the other' in engendered heritage representation and consumption. The chapter focuses upon a specific case study and builds on the work of Edensor and Kothari (1994) by extending their analysis of the 'masculinisation of Stirling's heritage' through a more detailed conceptual and empirical analysis of the ways in which heritage tourism is represented within the Scottish town of Stirling. Whilst the town's tourism industry has undoubtedly benefited in the 1990s from the films *Braveheart*, *The Bruce* and *Rob Roy*, it has relied upon particular forms of masculinism, militarism and nationalism to create gendered spaces, places and landscapes. These gendered places are created and contested through the representation and interpretation of the iconography of a heritage landscape which emphasises masculine visibility and superiority in landmarks and monuments, buildings and statues, signs, symbols and banners, as well as the content and discourse of postcards and promotional literature.

Following an introduction to heritage tourism and gendered representations of spaces, places and landscapes, this chapter examines the concepts of the 'tourist gaze' and 'othering' as significant means of creating and maintaining gendered space in heritage tourism. Ways of seeing and being are then examined in relation to the case study of Stirling. The analysis includes place promotion of the gendered tourism identity of Stirling, the male dominated landscape of Stirling, and the mythical representations employed in the creation of Stirling's gendered tourism spaces and places.

Gender and heritage tourism

> History is the remembered record of the past: heritage is a contemporary commodity purposefully created to satisfy contemporary consumption. One becomes the other through a process of commodification.
>
> (Ashworth 1994: 16)

Sophisticated marketing strategies are now used to market places like products to be consumed. Places have been constructed and marketed as attractive because they are different. Distinctive labels and images have been used to identify the uniqueness of a space and place to the tourist constantly in search of 'the Other'. The 'heritage industry' has resulted in the creation of 'a whole new breed of attractions and intermediaries who supply culture specifically for tourist consumption' (Richards 1996: 13). This new heritage industry has made connections between specific places and traditional industries in the case of Beamish in the former coal-mining area of north-east England.

The role of gender in the construction of these landscapes seems yet to capture the geographical imagination. Swain (1995: 248) emphasises that tourism,

> as a vehicle of economic development, is ripe for gender analysis, following the lead of the literature on gender and development. Furthermore, tourism practices reflect representational issues of identity and nationhood in the marketing and consuming process between hosts and guests.

So whilst tourism is frequently cited as 'the world's fastest growing industry' or 'the world's largest industry' it is less frequently identified as the world's most sex-segregated industry or the world's most sex-role stereotyped industry. Although there is a growing body of literature which analyses the gendered construction of employment and management within tourism, there is little research identifying the gendered representation, production and consumption of tourism. Similarly, travel and tourism have become associated with a glob-alised 'melting pot' where postmodern deconstruction and reconstruction have induced the breakdown of previous national, cultural and geographical bound-aries. But, paradoxically, the mechanisms employed in contemporary tourism development have simultaneously served to strengthen, rather than destabilise, gendered representations of space and place alongside notions of nationalism and bounded cultural identity:

> Tourism, as leisured travel and the industry that supports it, is built of human relations, and thus impacts and is impacted by global and local gender relations.
>
> (Swain 1995: 247)

A gender analysis of tourism therefore has to address issues related to the social construction of space and place in addition to expanding existing analyses of tourism provision, employment and management. In short, tourism needs to be considered not just as a type of business or management but as a powerful cultural form and process which both shapes and is shaped by gendered constructions of space, place, nation and culture. Rather than being a physical or objective reality, 'landscape is a cultural image, a pictorial way of representing, structuring or symbolizing surroundings' (Daniels and Cosgrove 1988: 1). Landscape is there-fore a social construction which represents and symbolises a social and cultural 'geography of the imagination':

> Whether written or painted, grown or built, a landscape's meanings draw on the cultural codes of the society for which it was made. These codes are embedded in social power structures, and theorization of the relationship between culture and society by these new cultural geo-graphers has so far drawn on the humanist Marxist tradition of Antonio Gramsci, Raymond Williams, E.P. Thompson and John Berger. All of these authors see the material and symbolic dimensions of the produc-tion and reproduction of society as inextricably intertwined.
>
> (Rose 1993: 89)

In the work of the humanist Marxist theorists cited by Rose, landscape is constructed as part of capitalist hegemony where patriarchy is included within an analysis of capitalist power structures. It could be argued, however, that heritage has the potential to represent the reproduction of a past patriarchal hegemony by a present engendered iconography, so that there is a two-fold displacement of women spanning history and heritage and merging fact with fiction.

In addition to reading written material as text, heritage landscapes too can be read as the subject of textual analysis, interpretation and deconstruction. This chapter utilises textual analysis in its focus upon the iconography of heritage production and representation rather than any examination of practices of heritage consumption and consumer behaviour. It is acknowledged, however, that the interface and interaction between places and people, objects and subjects, representations and interpretations provides the context for contestation and negotiation of the performance of gender and power.

There has been relatively little research examining the gendered nature of landscape production, interpretation and consumption and even less examining heritage landscapes, spaces and places. Edensor and Kothari (1994: 165) argue that heritage production and consumption is gendered through a series of processes which 'articulate masculinised notions of place and identity, and male dominated versions of the past which privilege white, male, heterosexual experience and activity'. In relation to Stirling, they go on to identify the three sites of the Argyll and Sutherland Highlanders Museum at Stirling Castle, the Bannockburn Monument, and the Wallace Monument as forming the 'Stirling Triangle', with each point reflecting 'particular and partial histories and myths, male-defined landscapes, and gendered national identities'. The tripartite message of nationalism, militarism and masculinism is reinforced by the addition of other male icons and visual imagery drawn from both fact and fiction, history and heritage.

Gender and the tourist gaze

Urry's reading of the 'tourist gaze' (1990) was influenced by the work of post-structuralist theorists such as Foucault (1976: 89). The theory that knowledge is both socially constructed and socially constructing is central to the concept of 'the gaze'. But the power of the gaze is unequally distributed and both the object and subject of the gaze are constructed according to the locus of power and control. As a heterogeneous group, tourists are differentially empowered in their gaze upon the landscape. They engage in a dialectic process of projection and reflection, bringing a diversity of meanings to the object of their gaze. Simultaneously, however, these meanings are informed and renegotiated by the dominant reading of parti-cular objects, sites or landscapes. As Bender emphasises:

> Landscapes are thus polysemic, and not so much artifact as in process
> of construction and reconstruction . . . The landscape is never inert,

people engage with it, re-work it, appropriate and contest it. It is part
of the way in which identities are created and disputed, whether as individual, group, or nation-state.

(Bender 1993: 3)

In this way it is quite conceivable that readings of monuments celebrating
Scottish nationalism have associations with class and ethnicity which may cut
across gender. Thus there may be greater commonality of reading between those
tourists identifying as Scottish nationalists of both sexes than between women
tourists from Scotland and England for example. Ashworth (1994: 15–16),
however, emphasises the role of individual agency in the construction of heritage
by stating,

> there is no national heritage product but an almost infinite variety of
> heritages, each created for the requirements of specific customer groups;
> viewed from the side of the customer, each individual necessarily deter-
> mines the constitution of each unique heritage product at the moment
> of consumption.

Ashworth's interpretations of the heritage producer and consumer are diamet-
rically opposed to those of Hewison (1987). Drawing on the work of the Frankfurt
School, Hewison focused on the hegemonic economic, social and cultural power
of the producer and the naïveté of the unquestioning and undifferentiated mass
consumer. Whilst a feminist analysis of Ashworth's work might draw attention
to the over-emphasis on individual agency, feminist critique of Hewison's work
might point to the lack of recognition of hegemonic masculinism as a dominant
power-broker in the representation and interpretation of heritage.

Boorstin (1964) offered a critique of tourism as a 'pseudo event' or as 'staged
authenticity' where tourists do not act as cultural dupes but question, contest
and reappraise the representations constructed by heritage intermediaries.
Evidence of such a dialectic between gendered power and resistance is clearly
demonstrated in the modification of one of Stirling's 'masculinised' heritage sites.
The inclusion, in 1997, of a display of statues of prominent Scottish women
within the Wallace Monument was in direct response to protests at an entirely
male exhibition. In spite of these contestations, however, it is difficult to dispute
the existence of dominant representations and readings of Stirling's heritage
where the tourist gaze is directed towards male icons and masculinist symbolism
embodying both nationalism and militarism. As Sharp (1996: 103) states:

> Gender cannot be teased out of other relations of power which consti-
> tute individual subjectivity but must be seen to exist contingently in all

situations – no one can be without gender and in most social locations this is a powerful aspect of subjectivity.

In relation to cultural tourism, Swain (1995: 249) asserts that:

> Host societies differentiated by race/ethnicity, colonial past, or social position from the consumer societies are sold with feminised images. The tourism product – as a combination of services, culture, and geographic location – is consumed in situ, in various transactions from tourist gazing to the selling of otherness.

But the gendered representation of tourism in Stirling is somewhat different to this model. Whilst notions of the male gaze and the sexual objectification of women are familiar, notions of the male gaze directed at particular forms of masculinity, rather than femininity, are less frequently discussed as mechanisms in the performance of gender and power. In Stirling's major heritage attractions, the male gaze is directed towards masculinist icons of a brutal and bloody past where representations of 'the gaze' and 'the other' are interwoven with constructions of women as the 'invisible other'.

Constructing the other and the invisible other

The creation of a unique place or tourist destination frequently employs the social construction of the other. This has been commented upon at length within the literature of tourism anthropology and tourism sociology and there is widespread agreement that tourists often engage in what Van den Berge (1994) has described as 'The Quest for the Other' (Urry 1990; MacCannell 1992). The reverse side of the coin in searching for the other, however, is searching for 'the self'. Whilst the search for the authentic other may lead to the discovery of the authentic self, Brown (1996: 38) has argued that 'the tourist seeks out the inauthentic Other in the quest for the authentic Self'. Brown's view provides a role for the tourist which builds upon both Boorstin's (1964) and MacCannell's (1992) critiques of tourism as a 'pseudo-event' or as 'staged authenticity' and gives recognition to tourists as a heterogeneous group with different 'cultural capital' (Bourdieu 1986).

The attempt to create and market an identity based on a unique heritage must coexist, however, with the need to facilitate the tourist's identification with that identity. Thus a balance must be sought between presenting something which is different and presenting something sufficiently familiar for the tourist to identify with. This can be achieved by using vehicles such as nationalism and militarism, combined with ancestral heritage, to provide the connections between the past and the present or between the familiar and 'the Other'. Most writers in tourism anthropology and tourism sociology cite Said (1978) as the originator of the concept of 'the Other' in reference to the Western social construction of 'Orientalism' (Selwyn 1996). The anthropology of tourism therefore discusses the concept of 'the Other' in relation to anthropological representations of race and ethnicity and this male-dominated discourse ignores the wealth of feminist writing on 'the Other' underpinned by the much earlier references to woman as 'the Other' by de Beauvoir (1949: 18):

> Humanity is male and man defines woman not in herself but as relative to him; she is not regarded as an autonomous being . . . She is simply what man decrees . . . She is defined and differentiated with reference to man and not he with reference to her; she is the incidental, the unessential as opposed to the essential. He is the Subject, he is the Absolute – she is the Other.

Rose (1995: 116) explains the process of 'othering' as 'defining where you belong through a contrast with other places, or who you are through a contrast with other people'. Such a process inevitably then defines the binary relations of norms and deviants, centres and margins, cores and peripheries, the powerful and the powerless. Rose (1996) later draws on the work of Irigaray to illustrate that the binary distinctions between self and other, and real and imagined space, are part of what Butler (1990: 13) has referred to as 'the epistemological, onto-logical and logical structures of a masculinist signifying economy'.

But whilst women are reified and objectified within sex tourism, we are frequently invisibilised within heritage tourism where emphases upon nation-alism and the construction of the nation-state draw upon a history which frequently renders women the invisible 'other'. Here women are absorbed, subsumed, and finally invisibilised by engendered representations of nation.

Constructing Stirling: gender and place identity

Sharp (1996: 97) comments that: 'National identity is a central aspect of contem-porary subjectivity and yet, certainly within the discipline of geography, its articulation with gender has largely been ignored.' She goes on to state that, 'The symbols of nationalism are not gender neutral but in enforcing a national norm, they implicitly construct a set of gendered norms' (1996: 98). Penrose (1993: 29) comments that the construction of the category 'nation' is depen-dent upon the social construction of people, places and a mystical bond between people and places and this chapter now goes on to illustrate the engendered nature of these relationships in the case of Stirling.

Like other tourist destinations, Stirling has attempted to create an identity of place which distinguishes the town and surrounding countryside from competing destinations. Traditionally, this identity has been constructed around the image of Stirling as 'The Gateway to the Highlands' but this identity implies a short-stay tourist destination through which visitors pass en-route to their final destination. In order to encourage longer stays and greater income generation, Loch Lomond, Stirling and the Trossachs Tourist Board, together with Stirling District Council attempted to market the area as 'Royal Stirling' and 'Historic Stirling' during the late 1980s and early 1990s. Promotional literature developed since the late 1980s emphasises the area's links with Robert the Bruce, William Wallace and Rob Roy MacGregor, alongside frequent references to Stirling Castle as a royal residence of the Stuart Kings. In 1995 Scotland's forty-two local tourist boards were rationalised into a system more akin to that of England and Wales where fewer, larger tourist boards attempt to achieve greater economies of scale. But the heterogeneity of the landscape and tourist attractions in and around Stirling has created difficulty in defining one overall identity or brand label which encapsulates the diversity of the product. The new tourist board has the rather unwieldy name of 'Argyll, the Isles, Loch Lomond, Stirling and Trossachs'.

The production of the Hollywood blockbuster, *Braveheart*, could not have been timed better to provide the new tourist board and the new local authority of Stirling Council, created in April 1996, with a clearer brand image which encapsulates elements of the history and geography of the area, together with particular notions of nationalism, militarism and masculinism, which appear to have captured the popular imagination following the success of the film. In the summer of 1996 the town experienced increased visitor numbers and the local tourist board launched a television marketing campaign utilising this brand imagery by emphasising 'Stirling Is Braveheart Country'. Ironically, and in contrast to recognised profiles of tourists and visitors to heritage sites, many of Stirling's new visitors are young Scottish men with a reasserted interest in Scottish nationalism motivated by the exploits of an Australian actor (Mel Gibson) filmed largely in Ireland.

The heritage landscape of Stirling is a product of both physical and cultural geography. Stirling is situated at the narrowest point between east and west in the whole of the British Isles and also overlooks the Carse of Forth which forms the divide between the Highlands of Scotland to the north and the Lowlands to the south. The town itself grew down the side of one of three volcanic outcrops in the area, which rose above the marshland of the Carse, a deep impenetrable moss which was not drained until the eighteenth century.

Stirling's economic and industrial history is an interwoven mix of market and garrison town supported by the primary industries of coal and agriculture. With the demise of coal mining, and its other related secondary industries in the 1980s, the town focused upon financial and retail services, its university, and tourism as its major sources of income. There appears to be a sense in which the town sought to replace the masculinity of the collapsed male dominated industries with the form of masculinity embodied in Stirling's nationalistic and militaristic history with William Wallace, Robert the Bruce, Rob Roy McGregor and the Stuart Kings featuring heavily in product development, place promotion and marketing literature.

A number of ambitious tourism plans were proposed in the mid-1980s by the then Labour controlled local authority under the umbrella of 'Futureworld', later modified. New developments have included extensive marketing of Stirling's nationalistic and militaristic history with William Wallace, Robert the Bruce, Rob Roy McGregor and the Stuart Kings forming the new iconography of the Stirling landscape.

Stirling: a male dominated landscape?

Tilley (1993) stresses the importance of the relationship between architecture and topographic form where the interaction between the built environment and the natural environment can create a specific sense of place and power. Bender (1993: 10), introducing Tilley's work, comments that,

> architecture acts as a lens for perceiving the landscape around monu-
> ments. Its relationship with the landscape may prove a powerful means
> for legitimizing power relations and naturalizing social order.

The representations of Wallace and Bruce within the town, as well as overlooking the town from the two surrounding hillsides, take a number of forms. Each male icon has a street and a number of public houses named after them and there are large statues of William Wallace at two of the most focal points within the town. The first is on the frontage of a public building called the Athenaeum which is situated at the top of King Street in the centre of the town and overlooking the main shopping area, and the other is less than one hundred metres away on the front of the Municipal Building (the town hall or local council offices). There is also a large statue of Robert the Bruce in front of the castle which overlooks the castle esplanade and the view over to the Wallace Monument some two miles away.

More recent representations have taken the form of symbols and logos, with Robert the Bruce representing the area of Bannockburn in a series of banners, again in King Street, and each identifying a different local area. The logo of 'Futureworld', a local tourism development and promotion project from the 1980s, consisted of Robert the Bruce on horseback surrounded by a rainbow.

'Key to the Kingdom' on the logo of Stirling Council could be seen as a reference to Stirling as the historic capital of Scotland, as the entrance to the Highlands, and as the locus of the history of the making of Scotland as a nation. In addition, postcards feature numerous images of the two icons from both near and far, conveying the heroic guardianship of both the town and the nation.

Stirling's male-dominated and male-defined history is evident in the names of buildings and their previous uses. Many of the buildings of the old town are now used as tourist attractions or as restaurants and coffee shops, primarily intended for tourists. In 1994 the Stirling Heritage Trail opened featuring many of the buildings of the old town but without any critique of the gendered nature of the town's heritage or of the representation of that heritage. Old town buildings named after their original male residents include John Cowane's House, Glenngarry Ludging, Spittal's House, Darnley's House, Norrie's House, Auchenbowie's House, Mar's Wark, the Argyll's Ludging, and the Erskine Church.

In addition to residential buildings and churches Stirling's new heritage trail, which starts with 'a soldier's view' from the castle esplanade, includes many other buildings with male histories; the Old Military Prison, the Old Grammar School for Boys, the Old High School, the Tolbooth and Prison, the Mercat Cross, and the Boy's Club. So whilst a tour of such heritage attractions might inform the tourist about some of their forefathers, there is no information about our foremothers because women are invisibilised from the local domestic history as well as from national political history and heritage.

The symbiotic relationship between masculinism and nationalism has been well documented (Yuval-Davis and Anthias 1989; Enloe 1989; and Walby 1992). Less frequently discussed, however, is the role that nationalism plays within tourism promotion (McCrone, Morris and Kiely 1995). A variety of forms of media have combined to reassert Scottish nationalism in recent years and films such as *Braveheart*, *The Bruce*, and *Rob Roy* have acted as catalysts for a tourism industry desperately in search of the familiar 'other'.

The Wallace Monument epitomises the inter-relationship between masculinity, militarism and nationalism in heritage production. The monument was refurbished to incorporate a variety of new visual displays in time to benefit from the additional tourists generated by the film *Braveheart*. The 220-foot high tower overlooks Stirling and the River Forth and was originally completed in 1869 as a monument to William Wallace who had led the defeat of Edward I of England at the Battle of Stirling Bridge in 1297.

Heritage production within the monument uses a powerful combination of written text, visual imagery and music to recreate the past. Pictures, statues, audio-visual displays, technology induced drama, and costume combine to create an environment where the visitor appropriates, contests, and negotiates engendered representations of heritage. Tourists entering the monument are overlooked by a huge bronze statue of Wallace with his sword drawn. The two-handed broadsword is then displayed on the first floor of the monument alongside a picture board display which presents a historical account of Wallace's life under titles like 'The Struggle Continues' and 'Capture and Execution'. In the centre of the same floor is a reconstruction of an English battle tent guarded by one of Edward's knights. Visitors are invited into the tent to witness 'a dramatised reconstruction and a talking head of William Wallace', who has been captured by the English following years of hiding after the Scottish defeat at the Battle of Falkirk in 1298.

Whilst the technology may be commendable, the text and discourse of Wallace's talking head require more critical analysis. Although in shackles, Wallace's posture and voice still command attention from his pulpit-like position as he gives his version of history. Wallace introduces his account by stating, 'Men must have their power. They seek to influence, to strengthen their position, to be seen as something in other men's eyes.' Wallace is constructed as the protector and voice of the nation where the nation is constructed as passive and feminine. Rose (1993) emphasises the feminisation of landscape as a means of portraying woman as nature and man as culture or woman as the passive background upon which man acts out the foreground. Here, there is a sense in which landscape becomes the nation and culture becomes nationalism; nationalism is the masculinised culture acted out upon the feminised backdrop of the nation.

On the second floor of the monument is the 'Hall of Heroes' where a plaque informs visitors that,

> In this vaulted chamber you'll meet other great Scots, sculpted in marble. Writers, explorers, inventors and statesmen are here, including King Robert the Bruce, Sir Walter Scott, David Livingston, Robert Burns and James Watt, among others.

The 'Hall of Heroes' is a celebration of the men and values of the enlightenment and, as McDowell (1993: 316) has commented, 'after an Enlightenment to themselves, it is hardly surprising that its creators are reluctant to relinquish the centre stage'.

The iconography of nationalism is inseparable from that of militarism in the representation of Stirling's heritage. Wallace and Bruce are seen as national heroes because of their military victories. Stirling Castle and its regimental museum is a tourist attraction largely because of its militaristic heritage. Complementing the statue of Robert the Bruce overlooking the castle esplanade is another statue of a soldier from the local regiment, the Argyll and Sutherland Highlanders, depicted with his bayonet drawn during the Boer War. There is a further sword-bearing statue in the form of Rob Roy MacGregor positioned at another focal point on the edge of the old town, and the Old Military Prison reinforces the militaristic theme and weaves further the web of masculinism, militarism and nationalism.

Stirling: masculinist myths as heritage?

Our violent past has been constructed as a tourist attraction of the present where spaces and places which witnessed previous atrocities are now commodified as contemporary commercial enterprises. The tourist industry has expanded its provision from war museums and battle-site visitor centres to re-enactments, living history dramatisations, and 'visitor experiences' as the postmodern tourist searches for greater novelty and virtual reality. We can now visit Auschwitz, Alcatraz, and 'Jack the Ripper' murder scenes where violence, imprisonment, genocide, murder and mutilation have become seen as legitimate tourist attractions.

Stirling's Old Military Prison opened in April 1996 as a combination of tourist attraction and office space for local businesses. The redevelopment took over three years and cost £2.6 million of public money with funding coming from Forth Valley Enterprise, Stirling District Council, and the European Regional Development Fund. The attraction is modelled, to a large extent, on Inverary Jail whose 1996 promotional literature stated that the attraction had won a number of awards and recommendations for its displays of 'torture, death and damnation: the story of Scottish crime and punishment 1500–1700' which features 'an introductory exhibition with blood curdling details of mediaeval punishments'. Visitors are introduced to life-like wax models of male prisoners and guards and an audio system conveys the sounds of prisoners in agony during and after torture. Neither the prison in Stirling nor in Inverary acknowledge the gendered representation of an equally gendered history.

Such representations of history result in a heritage which appears to embrace forms of masculinist myth making in an attempt to promote tourism. Thus Rob Roy MacGregor has had a heroic past constructed for him through postcards, statues and a visitor centre where tourists are encouraged to make up their own minds as to whether Rob Roy was a 'Hero or Villain?'. The language used to describe the less than heroic aspects of his life and character provides few challenges to the type of masculinism, violence and nationalism which are frequently constructed through a romanticised heritage. Visitors to the Rob Roy and Trossachs visitor centre are invited to contemplate Rob Roy as a 'rogue' and a 'ruffian', words which fail to convey the full horror of the violence prevalent at the time and which also serve to strengthen the imagery of male violence masquerading as emancipatory nationalism.

Tourism development in Stirling has also focused upon Ghost Walks which are evening guided tours on foot with a number of additional costumed characters appearing en route. These tours provide an even clearer example of the way in which tourists are invited to empathise with particular forms of masculinity and violence. The major Ghost Walk starts and finishes with speeches by Allan Mair, the last person to be publicly hanged in Stirling. Mair was hanged in 1843 for the murder of his wife and his ghost is said to haunt the Tolbooth, now a restaurant. Although much of the information presented during the actor's performance is historically accurate, dramatic licence has allowed the performers to construct a pantomime-like scene where Allan Mair protests his innocence to the assembled audience, again in a 'roguish' manner, and the crowd is finally encouraged to voice their belief that Mair must be innocent of the crime for which he is about to be hanged.

The Ghost Walks, as a form of nocturnal street theatre, build on their daytime and early evening equivalents where a variety of theatre productions celebrating the life of William Wallace have been staged alongside other productions celebrating particular forms of masculinism, militarism and nationalism. In 1997 there were events and celebrations held throughout the town to mark the 500th anniversary of the Battle of Stirling Bridge in which William Wallace defeated the English. These events also coincided with the Scottish 'Yes' vote for devolution and provided an emphatic backdrop against which to stage productions with titles like 'To wear the white cockade'. These performances serve to promote tourism's permeable boundaries between myth and reality, fiction and fact, heritage and history. But as Kappeler (1986: 3) has stated:

> Representations are not just a matter of mirrors, reflections, key-holes. Somebody is making them and somebody is looking at them, through a complex array of means and conventions. Nor do representations simply exist on canvas, in books, on photographic paper or on screens: they have a continued existence in reality as objects of exchange; they have a genesis in material production. They are more 'real' than the reality they are said to reflect. All of these factors somehow straddle the commonsense divide between fiction and fact, fantasy and reality.

Conclusion

This chapter has sought to contextualise and engender heritage by teasing out the multiplicity of interconnections between space, place and gender experienced in relation to the social construction of heritage landscapes and iconographies. Constructions of 'the gaze' and 'the other' were situated as central concepts in the process of heritage representation and consumption. Stirling then acted as a case study which illustrated the embodied nature of 'the gaze' and 'the other' within the interconnected mechanisms of place promotion, landscape construction, and mythical representation. Our engendered heritage has been identified as a complex cultural product which is negotiated, contested and, ultimately, has the potential for transformation of space, place and people. As Duncan (1996: 246) has stated, 'Rather than claiming space for a group, as in territorially based politics, moveable sites of resistance against exclusionary practices can break open such performances of power'. In conclusion then, there is no attempt here to reclaim space in a physical sense but to problematise the cultural construction of heritage landscapes as engendered and contextualised performances of power.

References

Aitchison, C. and Jordan, F. (1998) *Gender, Space and Identity: leisure, culture and commerce*, Leisure Studies Association: Eastbourne.

Ashworth, G.J. (1994) 'From history to heritage – from heritage to identity: in search of concepts and models' in Ashworth, G.J. and Larkham, P.J. (eds) *Building a New Heritage: tourism, culture and identity*, Routledge: London.

De Beauvoir, S. (1949) *The Second Sex*, Penguin: Harmondsworth.

Bender, B. (Ed) (1993) *Landscape: politics and perspectives*, Berg: Oxford.

Boorstin, D. (1964) *The Image: a guide to pseudo-events in America*, Harper and Row: New York.

Bourdieu, P. (1986) *Distinction: a social critique of the judgment of taste*, Routledge: London.

Brown, D. (1996) 'Genuine fakes' in Selwyn, T. (ed.) *The Tourist Image: myths and myth making in tourism*, Wiley: London.

Butler, J. (1990) *Gender Trouble: feminism and the subversion of identity*, Routledge: London.

Daniels, S. and Cosgrove, D. (eds) (1988) *The Iconography of Landscape: essays on the symbolic representation, design and use of past environments*, Cambridge University Press: Cambridge.

Duncan, N. (ed.) (1996) *BodySpace: destabilizing geographies of gender and sexuality*, Routledge: London.

Edensor, T. and Kothari, U. (1994) 'The masculinisation of Stirling's heritage' in Kinnaird, V. and Hall, D. (eds) *Tourism: a gender analysis*, Wiley: London.

Enloe, C. (1989) *Bananas, Beaches and Bases: making feminist sense of international politics*, Pandora: London.

Foucault, M. (1976) *The Birth of the Clinic*, Tavistock: London.

Hewison, R. (1987) *The Heritage Industry: Britain in a climate of decline*, Methuen: London.

Kappeler, S. (1986) *The Pornography of Representation*, Polity: Cambridge.

MacCannell, D. (1992) *Empty Meeting Grounds*, Routledge: London.

McCrone, D., Morris, A. and Kiely, R. (1995) *Scotland – The Brand: the making of Scottish heritage*, Edinburgh University Press: Edinburgh.

McDowell, L. (1993) 'Space, place and gender relations: Part II. Identity, difference, feminist geometries and geographies', *Progress in Human Geography* 17(3): 305–18.

McKean, C. (1985) *Stirling and the Trossachs*, Edinburgh: Royal Incorporation of Architects in Scotland/Scottish Academic Press.

Penrose, J. (1993) 'Reification in the name of change: the impact of nationalism on social constructions of nation, people and place in Scotland and the United Kingdom' in Jackson, P. and Penrose, J. (eds) *Constructions of Race, Place and Nation*, UCL Press: London.

Richards, G. (ed.) (1996) *Cultural Tourism in Europe*, CAB International: Wallingford.

Rose, G. (1993) *Feminism and Geography*, Polity: London.

Said, E. (1978) *Orientalism*, Routledge: London.

Selwyn, T. (1996) *The Tourist Image: myths and myth making in tourism*, Wiley: London.

Sharp, J. (1996) 'Gendering nationhood: a feminist engagement with national identity' in Duncan, N. (ed.) *BodySpace: destabilizing geographies of gender and sexuality*, Routledge: London.

Swain, M. (1995) *'Gender in Tourism', Annals of Tourism Research* 22(2): 247–66.

Tilley, C. (1993) 'Art, architecture, landscape (Neolithic Sweden)' in Bender, B. (ed.) *Landscape: politics and perspectives*, Berg: Oxford, 139–68.

Urry, J. (1990) *The Tourist Gaze*, Sage: London.

Van Den Berghe, P. (1994) *The Quest for the Other*, University of Washington: Seattle.

Walby, S. (1992) 'Women and nation', *International Journal of Comparative Sociology* 33(1): 81–99.

Watson, B. and Scraton, S. (1998) 'Gendered cities: women and public leisure space in the "postmodern city"', *Leisure Studies* 17(2): 123–37.

Yuval-Davis, N. and Anthias, F. (1989) *Women Nation-State*, Macmillan: London.

6

HYPER-REALITY IN THE OFFICIAL (RE)CONSTRUCTION OF LEISURE SITES

The case of rambling

Neil Ravenscroft

Introduction

Rambling remains one of the most popular leisure practices undertaken in the countryside of England and Wales (Spink 1994). According to the Countryside Commission [CC], some 14 per cent of the 50 million annual visitors to the countryside cite long walks as the principal reason for their visit (Countryside Commission 1992, 1995a). In the main, these walks are informal and have traditionally coexisted with other uses of land, occupying a mixture of legal rights (such as the public rights of way network), subsidised provision (such as country parks) and *de facto* and licenced access to both public and private land.

While for the most part the social demarcation between owner and non-owner may be evident, the consumption claims made by walkers have not generally been of sufficient force to warrant concerted opposition from landowners and occupiers. Where conflicts of use have emerged, the powers of exclusion associated with property have been brought into play. These powers have, for the last 50 years, been supplemented and underwritten by a statutory planning system in which most non-agricultural uses of land have effectively been subject to control and, in the countryside, exclusion (Ravenscroft 1992; Spink 1994; Ravenscroft and Reeves 1997).

Despite the opportunities offered by the various means of access to the countryside, a long-standing issue has been the extent to which ramblers can gain greater accessibility, particularly to uplands and open countryside (Ramblers' Association 1984; Countryside Commission 1991; Blunden and Curry 1989). This claim for the right – or freedom – to roam has had dimensions of both practicality (gaining access to new and remote environments) and ideology

(Shoard 1987). It has also generated some notable confrontations, particularly the inter-war 'mass trespass' of Kinder Scout in the Peak District and more recently the trespasses in the 'Forbidden Britain' campaigns.

Official response to these demands has been severe, ranging from imprisonment for those held responsible for the Kinder Scout trespass, to the criminalisation of some forms of access-related trespass under the Criminal Justice and Public Order Act 1995. When further threatened, as has been the case with the recent publication of the Government's consultation paper on freedom to roam (Department of the Environment, Transport and the Regions [DETR] 1998) some landowners have withdrawn access altogether, as the recent case of the closure of Ranmore Common in Surrey indicates (Hetherington 1998).

In contrast recent governments have also sought to increase the accessibility of the countryside, both through improvements in current provision (Countryside Commission 1993, 1996; Parker 1996; Curry and Ravenscroft 1997) and in tacit recognition of the claims for greater freedom to roam, through new forms of provision, linked to the reform of European agricultural policy. It is this new form of provision, granting either open or linear access over certain land within one of the EU-recognised agri-environment schemes (Ministry of Agriculture, Fisheries and Food [MAFF] 1993), which has generated considerable debate and, recently, deconstruction (Ravenscroft 1998).

In offering 'open' access, these sites are posited as replicas of pre-enclosure, pre-industrial rural Britain, with people free to wander at will over open countryside (although there is no overt suggestion that people undertook, or had the opportunity to undertake, such *leisure practices*). Yet, even in their physical sense, these sites do not form, in any part of the country, a continuous presence, while landowners and farmers have, in any event, prevented widescale dissemination of their location. Furthermore, as Werlen (1993) argues, it is not the spaces themselves which give meaning to practice, but practice – in this case *leisure practice* – which gives meaning to space. Thus, even if the location of the sites were known, mere physical simulation – the creation of the space – would not, in the absence of practice, create 'new' leisure sites, whether real or simulated.

As a result, it is apparent that these spaces have only an ironic, nostalgic, physical connection to the originals they are supposed to replicate, instead having more in common with Plato's simulacrum (the copy for which there is no original). In decentring the relevance of practice to the experience of leisure, these sites thus represent, in Baudrillard's terms, the 'perfect' simulatory machine, possessing a stability so clearly lacking in the 'vicissitudes of the original' (1994: 2).

This simulatory machine does not reflect the knowing irony often associated with the replication of originals (Urry 1990; Couldry 1998). Rather, there is a strong theme of policy which suggests that the hyperreality of these access sites is overtly 'created' (dissimulated), with this dissimulation used to mask a chasm between the message (freedom to roam) and the underlying ideological project (maintaining the hegemony of rural property ownership). Crucially, therefore, this policy theme seeks to play down the identity of the owner/farmer, still the focus of the rural policy community throughout Europe (Ravenscroft *et al.* 1998), in favour of an overt emphasis on the physicality of space itself.

In exploring these connections, this chapter is concerned with the ways in which the demands, however rhetorical, for particular forms of leisure practice have been appropriated to underpin the hegemonic project associated with the ownership and control of the countryside in England and Wales. In particular, it examines the ways in which claims and counter-claims about the significance of the relationship between people and environment (played out through leisure *practice*) have become a metaphor for the cultural contestation of values (Myerson and Rydin 1996) and, consequently, a formal element in the continuing hegemony of property (capital).

The chapter seeks to analyse the nature of the revised representations of informal leisure spaces in the countryside, considering how public demands for access to the countryside have been reinterpreted to facilitate and legitimate new forms of rural commodification and, thus, financial support for landowners and farmers. In so doing it is argued, reflecting Urry's (1995) contention about the relationship between consumption and place, that through the deceptive ideological vacuum of dissimulation, these sites effectively consume the identities of both landowners and recreationists (becoming, in effect, *all-consuming* places).

At the core of this deception, and linked to the *all-consuming* hyperreality of the sites themselves, there is a further – and fundamental – simulacraic transformation, in the recasting of the dominant (landowner) interests, towards a model of benevolent tutelage (Donzelot 1980). Under this transformation, landowners are constructed as willing 'voluntarily' to forego some of their libertarian freedom for the greater good of the general citizenry. As part of this voluntarism, landowners are seen to offer the attraction of a 'safe', managed countryside, in which the conduct and behaviour of 'unskilled' visitors can be guided and tutored. While offering a sanitised, hyperreal, construction of landownership, as socially responsible stewardship, this very process plays an active role in the reconstruction of citizenship. Rather than gain the new rights rhetorically offered by and through these sites the identity of the citizen is effectively consumed within the 'moral superstructure' which they have endorsed through their claims for, if not their actual use of, the sites themselves.

Countryside recreation and the hyperreal leisure site

Data from the Countryside Commission (1995a) indicate strongly that the most favoured outdoor leisure practices remain associated with activities such as rambling and walking. The core provision for informal access to the countryside of England and Wales is the 120,000-mile public rights of way network, over which there is a legal right of access on foot to all paths and a similar right over bridleways (about ten per cent of the system) for horse and bicycle riders. Although not evenly distributed around the country, with a predominance of paths in the more populated rural areas of the south and west, the system is generally well maintained and accessible (CC 1996; Ravenscroft *et al.* 1996).

In addition to legal rights of way there is a range of subsidised provision, small country parks and picnic sites usually operated by local government, and by a number of private landowners. Although car parking fees may be charged at these sites, people are generally free to wander at will. The National Trust, under its bylaw powers, allows free access to most of its 500,000 acres of open land, as does the Forestry Commission and a number of the water utility companies. Some of this access may include common land, although there is no general right of access to rural commons. A number of private landowners, particularly in the Peak District National Park, have negotiated access agreements under the National Parks and Access to the Countryside Act 1949, with local planning authorities, allowing open access to otherwise private upland areas close to the major northern conurbations. Finally, there is a class of *de facto* access to both public and private land where, although there are no legal or contractual rights, people have walked without hindrance from the owner (see Curry 1994; National Trust Access Review Working Party 1995).

Although outwardly reflecting a wide range of access *opportunities*, there is evidence that the total area of accessible countryside has been declining in recent years (Hetherington 1998). In particular, the sale of Forestry Commission and former public utility land has led to access areas being closed, while a recent (unreported) High Court decision has cast doubt on the legal accessibility of common land, even where such use can be shown to have occurred for over 20 years.

However, with an estimate that only one-eighth of the population is able to read a map (Glyptis 1995), research by Kay and Moxham (1996) indicates that little credence is apparently given to the needs and desires of users in the provision of this access to leisure spaces in the countryside. In part, this problem stems from a continuing lack of official recognition that informal recreation can have different meanings for different people. This is well illustrated in a recent discussion paper from the Countryside Commission (1995), where the claims made for the relationship between the quality of countryside and people's quality of life have not been subjected to any independent (nor, apparently, internal) analysis.

While unwilling to countenance freedom to roam having legal precedence over the individual rights of landowners, the 1992–97 Government was able to facilitate an apparently similar form of open access as part of an agricultural support package for farmers (DOE/MAFF 1995). Central to this incentive system is a two-tier designation of farmland, the most environmentally significant areas being accorded Environmentally Sensitive Area (ESA) status and financial support, while the remaining countryside can gain financial support through the Countryside Stewardship Scheme (CSS). Farmer reception to the designation of ESAs has been favourable; nearly 90 per cent of the area targeted was entered into management schemes (Whitby and Lowe 1994). The smaller CSS was similarly well received with nearly 5,000 agreements covering about 100,000 hectares of land after four years (Davies 1995).

A significant element of both schemes, recreational access, has been more thoroughly promoted in the CSS, mainly due to its antecedence as a Countryside Commission pilot scheme (ESAs having been developed by MAFF). Introduced in 1991, it sought to demonstrate that conservation and public enjoyment could be combined successfully with commercial farming (Countryside Commission 1994: 2). The agreements, in the form of contracts with farmers and others managing land, are for a fixed period of 10 years (Countryside Commission 1991a). Having been deemed a success by the government (DOE and MAFF 1995), the Scheme has recently been transferred, under the Environment Act 1995, from its pilot status at the Countryside Commission to a permanent status at the MAFF, as the primary incentive scheme for those farming outside ESAs.

Although central to the concept of CSS, public access is not a prerequisite for every site. Indeed, the Countryside Commission has stressed the point that not all land is suitable, either due to its location or to the presence of rare fauna or flora. Take-up of the public access option has been partial, leading supporters to defend the Scheme on the grounds that it conforms to the rigours of market allocation, in that the purchaser is under no obligation to purchase and any payment made is for the 'product' rather than simply the process, unlike other incentive schemes.

> The schemes allow farmers to identify relevant environmental services
> and goods which they can provide and the opportunity to market these
> and promote their role as custodians and managers of the countryside.
>
> (Bishop and Phillips 1993: 335)

There is really no attempt to deconstruct the identity of the 'purchaser' necessary for the market to operate. Indeed, references made by Bishop and Phillips to the 'public sector' as purchaser presumably ignore any aspect of accountability, given that evidence exists of public antipathy to paying farmers for access agreements (Clark *et al.* 1994, 1994a).

However, it is clear that not all land entered into Stewardship is subject to access agreements. The Countryside Commission estimate is that no more than one-third of the 2,963 agreements entered into in the first two years of the scheme contained provisions for access (Countryside Commission 1994), while the Country Landowners' Association has suggested that less than one-fifth of the total area in the Scheme is subject to access agreements (Bosanquet 1995). Yet no evidence indicates that this disparity results from the 'consumer choice' with substantial numbers of farmer or land owner applicants being denied the access payments.

Not all of those sites with access payments are suitable for their purpose, either because they are not part of the wider network of access to the countryside, or because they are physically inaccessible (Ramblers' Association 1993, 1994) Some CSS access sites do not represent new access at all, rather new agreements over existing *de facto* access sites (Ravenscroft *et al*. 1995). This hardly suggests the 'rules of the market' in evidence, let alone operation.

To qualify for CSS, access had to be an addition to what was previously available, even if the existing access was not through legal right. In its first year new public access was limited to open access sites (where the public is free to roam). Since then a linear option has been introduced, covering footpaths, bridleways and routes for the disabled. It has been estimated by the Countryside Commission that, by the end of the pilot scheme, linear access accounted for about half of all new access agreements.

Since access was supposed to be an addition to what was already available, no plans or representations of the access sites were readily available at the start of the scheme. As a consequence, agreements were reached with, and payments made to, farmers prior to the public having any knowledge of where they could walk (Ramblers' Association 1993). Initially, therefore, access sites were invisible, if not also abstract, with no impact as potential sites of leisure practice. Subsequently plans have been produced, providing a hitherto unavailable reality – but only for those knowing where to obtain a copy. Any wider circulation of information about the sites, especially by adding them to Ordnance Survey maps, has been resisted.

> Access marked on an Ordnance Survey map may no longer be available in ten years' time . . . but someone walking there could still produce the map and say, 'Here it is, it's marked down'; that cannot take any account of the fact that circumstances change, so that there is a risk there.
>
> (Bosanquet 1995: para. 869)

Thus few plans are in circulation and, at the end of the agreement, they can and will, presumably, be withdrawn, so effectively denying the reality and delegitimising the site.

In its recently published consultation document earlier Labour Party commitment to legal reform (1997) is downplayed, being replaced by the 'principle'
that people should have greater freedom to explore open countryside (DETR
1998, 1998a), positioning landowners – rather than the law – as the principal
agents in the supply of access.[1]

While the demise of Labour's commitment to freedom to roam is likely to
generate much dissent (Rowan-Robinson 1998; Wilmore 1998), it would suggest
that access policy is set to prolong the schism between the message and the
reality. While continuing to proclaim an egalitarian emphasis on accessibility,
therefore, Labour's policy, like that of all preceding governments, is much more
centrally associated with underwriting landowner rights, particularly in maintaining their ultimate right to exclude those seeking access, if the terms upon
which they do so are not to the landowner's liking.

Countryside as place and space

In attempting to deconstruct the nature and significance of these new paid
access sites, it is clear that an issue of central concern is the relationship between
the sites as physical spaces and their meanings as places where a multi-
textuality of leisure practices can occur. Countryside, in any context, is neither
solely – nor simply – a space, assuming as it does a particular semiotic
significance in late modernity (Halfacree 1993). Rather, it comprises a number
of identities beyond its physical form, being both a product presented for
consumption (its social construction and representation), as well as a place
where many different practices occur: Lefebvre's (1991) distinction between
the 'conceived' representation of space and the 'lived' space of representation.

The essential problematic of the countryside is, thus, that causal spatial determinism becomes juxtaposed with what Keith and Pile (1993: 25) describe as
the 'malleability of the symbolic role of landscape'. Rather than seeing
the countryside purely, even predominantly, as a site or place where practices
and processes happen *for all sorts of reasons*, it is all too often appropriated as a
signification of certain social or cultural values – *it is reified as both product and
location*.

Consequently, agencies like the Countryside Commission continue to see the
countryside in incontestable terms, while utilising highly contestable notions such
as 'quality' and 'our finest countryside' (1995) to describe it and, in the process,
maintain stasis. Stasis in this case involves the retention of existing land use,
tenure and ownership patterns, both denying any room for alternatives as well
as ensuring that outsiders, the 'non-landed', are constructed as 'other', requiring
both tutelage and management:

People are more able to enjoy the countryside . . . when they have the knowledge and confidence to explore it for themselves.

(Countryside Commission 1995: 17)

Yet space is neither an object nor *a priori*, but a frame of reference for actions (Werlen 1993). Since the social world is produced not by space but by social actions and practices, it is these actions and practices which must constitute it. Accordingly space, such as the paid access sites, cannot be empirical in a deterministic sense, but can only be a classificatory concept within which the problems and possibilities of practice can be examined (Werlen 1993).

Rather than being causal therefore, space – the countryside – is fundamentally political (Massey 1993). That is, it is not so much space which determines practice, but practice which fills, and gives meaning to, space. Consequently, rather than the physical representation of space (the actual access sites) being the source of power which dominant interests wish to protect, it is the possibility and actuality of practice which generates that power. Space, or territoriality, thus becomes the metaphorical arena in which the effects of power can be analysed and determined (Foucault 1980), but according to the uses of that space rather than the space itself.

The deceptive notions of social process underpinning the provision of the paid access sites is a prime example of the exercise of this power relationship. At the centre of the political debate have been arguments about the accessibility, in legal and social terms, of the countryside (Newby 1986). While it is argued that walkers have never enjoyed a greater range of opportunities for legal access (RICS 1989), it is apparent that in most respects the actual experience has declined (Bonyhady 1987; Hetherington 1998).

Although it is convenient to blame the enclosure of land in a spatially deterministic sense for the growing dissociation between ownership and usage, the issue is more cultural and contextual. For rather than the regulation of access being of incidental concern to the ownership and management of land, it has been central in its legitimation (Ravenscroft 1995). Public demand to recreate on private land is the very manifestation of Foucault's power metaphor, with a level of accessibility 'protected', as an allowance or dispensation, for those 'unsuited' to the ownership of rural property (Thompson 1993).

Consequently, the legal construction of countryside recreation has been much more closely associated with a 'civilising' mission than with the narrower legal definition of rights or 'freedom' of access. Constructing those demanding greater rights of access to private space as 'other', what the rural landed are not, the power of the territoriality implicit in the tenure structure effectively decentres the debate from questions about the power and signification of ownership to those about the nature and use of space. As Baudrillard (1981) suggests, this signification centres on new forms of power and discrimination which go beyond mere possession, into social organisation and usage. Where once it was sufficient for landowners and farmers simply to own land, therefore, it is now necessary for them to be able to demonstrate their power of control over it.

In common with the 'unpeopled' spatial determinism of the Countryside Commission, landowners and farmers have been forceful in attempting to construct the access issue as technical rather than social. They claim to be able to accommodate visitors to 'their' land through management rather than as a result of 'uncontrolled' free access (Bosanquet 1995). This suggests a politically-legitimised appropriation of power relations, with property ownership and management posited as a responsibility unsuited to the evident ephemerality of 'ordinary' people: they may visit the countryside, but only in designated areas, under supervision and solely to participate in 'approved' activities (Ravenscroft 1995).

The issue of access to the countryside further demonstrates the centrality of leisure and leisure practices to questions of space. While countryside sites can feature a multiplicity of uses, it is only really in the case of leisure that the uses or practices are beyond the direct control of the owner or manager. Equally, countryside sites are viewed, in a deterministic sense, as unique in leisure terms, thereby embuing them with a semiotic meaning lacking from other sites (Crouch and Ravenscroft 1995). The separation of the leisure practice from the site of consumption is, therefore, essentially concerned with what Lefebvre (1991) sees as the shift from a construct of absolute space to that of abstract space. In the former the characteristics of the space assume a significance which is independent of its use – the country park, for example. The latter, abstract space, on the other hand is, theoretically at least, capable of being 'emptied', of deriving its meaning from the practices which it accommodates. In Warren's terms, such spaces reflect 'modes of representation that permeate virtually all landscapes and hence are inseparable from daily life' (1993: 173).

It is this construct of abstract space which is at the centre of a new reading of leisure in the countryside. For rather than deterministic ideas of the territory preceding the map and the manager, the opposite is increasingly the case, as Baudrillard suggests:

> Abstraction today is no longer that of the map, the double, the mirror or the concept. Simulation is no longer that of the territory, a referential being or a substance. It is the generation by models of a real without origin or reality: a hyperreal. The territory no longer precedes the map, nor survives it. Henceforth, it is the map that precedes the territory – *precession of the simulacra* – it is the map that engenders the territory . . .
>
> (1988: 166)

It is the notion of the simulacrum which explains the nature and form of paid access sites, creating models of realities that never existed, designed to be consumed by largely 'uncritical' consumers. Rather than an overt simulation of the 'freedom' of access to pre-enclosure Britain, however, the irony of paid access sites is that *they do not seek to simulate anything*, but to obscure an underlying message of social closure. This, in Baudrillard's (1994) terms, is dissimulation – pretending not to have what one has. The 'what' in this case is the social approbation of the rights of exclusion and closure associated with private property ownership and inherent in the 'voluntary' sale of access to the sites. Baudrillard analyses this use of the simulacrum for the purposes of deception, commenting that:

> The transition from signs that dissimulate something to signs which dissimulate that there is nothing marks a decisive turning point.
>
> (1994: 6)

The 'turning point' is, effectively, the very denial of ideology. In the former case – attempting to dissimulate something – there is an inherent ideological debate about the nature and construction of reality and imagery. This engages both with the extent to which paid access sites simulate pre-enclosure Britain, and the extent to which they actually represent increased opportunities for leisure practices. In the latter case – dissimulation that there is nothing – any pretext of ideology (signified by false representation) is removed, to '(conceal) the fact that the real is no longer real' (Baudrillard 1994: 13). In this case, the 'real' is the notion that there is still a moral superstructure through which the nature and extent of property rights can be debated and negotiated.

Compliance with the imperative of 'paid voluntarism' has, therefore, effectively delegitimated the rights claims of political citizenship, in favour of a commodified parody which has supplanted morality and ethics for the logic of capital and the market. Evidence of the success of this project lies not so much in the rhetoric of either the last Conservative administration, nor that of landowners and farmers, but in the apparent emasculation of the new Labour administration's commitment to secure freedom to roam over open country. Thinly disguised by a now-familiar rhetoric of pragmatism, the recent consultation paper (DETR 1998) endorses the ideological denial implicit in paid voluntarism, by conflating it (or a renegotiated version of it) with the undeniably political nature of the access *rights* which it originally sought to secure.

Conclusions

In essence paid access sites such as those in the Countryside Stewardship Scheme feature, in de Certeau's (1984) terms, a differentiation between the absolutist spatial determinism of the suppliers ('setting up' the site as space) and poten-tialities of practice which this offers (the transformation of space to place). However, given the level of dissimulation between the rhetoric and actuality of this interface, the physical, social and legal dimensions of the space are suffi-ciently clouded by abstract hyperreality to render irrelevant the practices and imagination necessary for the transformation to occur. The suppliers thus claim to be creating, through an ideologically neutral process of management, new sites which are both delineated on a plan and are legally accessible. The poten-tial user, on the other hand, sees none of this. Consulting Ordnance Survey Maps will not yield the information; nowhere is it made clear what is either allowed or expected of them should they happen across this land.

Although not an immediately obvious subject for an exploration of hyper-reality, given its modern or even pre-modern associations, it is clear that the production of government-funded access to private land in the countryside is precisely this — hyperreal. That the sites exist at all within the lives of most cit-izens must be open to doubt; only the most determined and informed are likely either to come across them or know what to do if confronted by them. They are equally clearly posited as imitations of originals which never existed — simulacra defined only according to plans — ephemeral diversions from the wider and deeper social and cultural process of continued wealth and power redistri-bution.

Indeed, it is arguably power which is really at the core of the process. For what is apparent is that the construction of the 'problem' and, hence, the solu-tion, is very much determined by the relative power of the parties to the debate. The dominant construction of this power is in terms of 'values'; that rather than relating to differential power itself, differences can be explained by reference to relative values, which themselves exist to be contested (Myerson and Rydin 1996). The dominant message is, therefore, that paid access sites reflect a 'mute ground' (Fitzpatrick 1992), upon which the possibilities of leisure practice, as an expression of a cultural and social dynamic, can — hypothetically — be subjected to values-based negotiation and settlement.

The discourse on values is far from neutral. Values have little credibility without the means of legitimation. It may suit certain groups or associations to construct relative positions in terms of values, but the relativity is more associ-ated with the power to substantiate those value claims than it is with values *per se*. It is clear that the dominant values associated with 'paid voluntarism' have little to do with the abstract and social construction of space and everything to do with the continuation of the spatial determinism upon which land tenure is based.

Any representation of the access 'problem' as one of values, or even rights, is, then, an obfuscation of what is more fundamentally a mechanism of alienation and control. This is based on deceptive notions of social process promulgated by the dominant ideological forces associated with property. At the core of this promulgation is the denial of the political nature of land tenure in favour of a form of hegemonised morality which privileges land-based rights over those of others. This dominance has even reached the stage of achieving the acquiescence of the wider citizenry in the deception that access claims have been materially substantiated by schemes such as Countryside Stewardship (see, for example, RICS 1989).

Rather than 'freedom to roam' or some similar evocation, what the CSS and similar schemes have promoted is the reinvention of the countryside as a series of unified, sanitised, culturally 'safe' sites where recreation is allowed or even encouraged, where people's respect for the countryside can be developed. Rather than a recognition of abstract space, however, this is more a corollary of past managerialism, with specific sites or attributes of land 'sacrificed' to the needs of leisure, as a diversion from more 'valuable' or 'high quality' land elsewhere. Now, however, the diversion is not from other land, but from wider and more deeply philosophical issues about the continuing spatial determinism of land ownership and management.

Rather than being related to the former liberal paradigm of citizenship, as the evocation of the right to roam suggests, therefore, these new leisure spaces represent, or at least reflect, a stage in the development of a form of 'post-citizenship'. Urry (1995) characterises this development as a shift from a political construct of citizenship based on residence to a privatised one based on the effective 'consumption' of rights. Whereas recreational access to private land was once based on 'presumed' rights associated with locality Urry argues that these presumed rights now increasingly require protection through acquisition.

Rather than submit the purchase of these rights to the open market, however, paid access schemes such as Countryside Stewardship seem to represent an apparent attempt by government to 'downplay' the commodification, in favour of a superficial parody of the former liberal 'rights'. This is by producing leisure spaces with similar physical form, but entirely separate cultural identity to those being denied in the wider context. This construct is underpinned by a shift in the nature of those being protected. In the place of the former protection of citizens through customary attitudes to simple trespass, for example (Bonyhady 1987), there is now the overt recognition of farmers' legal interests through direct financial support, coupled with the criminalisation of trespass in certain defined cases. This has led Marsden *et al.* to comment that:

> The most compelling message to emerge is the ability of landed interests to alter the representations of their position to take account of

changing circumstances, and to do so effectively in a context of declining economic and political power.

(1993: 97)

As a result, these leisure sites assume the identity, in Urry's (1995) terms, of the *all-consuming* place. For, in legitimising the dubious claims about new forms of demand-driven access provision, society is effectively – and actively – substantiating and underpinning the 'market as citizenship' metaphor. In place of any notion of 'citizen rights' being beyond financial equivalence, therefore, these sites are both commodified in themselves and commodifying of those who use, or seek to use, them. Rather than presenting new spaces capable of transformation through the practices and imaginations of those who visit them they effectively render meaningless the very notion of *leisure practice* as a self-generated form of place-related expression:

> A place . . . is the order . . . in accord with which elements are distributed in relationships of coexistence. It thus excludes the possibility of two things being in the same location (place). The law of the 'proper' rules in the place: the elements taken into consideration are beside one another, each situated in its own 'proper' location, a location it defines. A place is thus an instantaneous configuration of positions. It implies an indication of stability.
>
> (de Certeau 1984: 117)

Place as formulated in contradistinction to the physical space of paid access sites displays none of these attributes. It implies neither order nor stability. Rather, it is fully part of the process of dissimulation, of creating 'something' to obscure 'nothing'. By engaging in the legitimation of these sites, through practice and imagination, the transformation of space to place is no more than an illusion, created by and for the continuance of the existing dominant ideology.

Notes

1 Contrary to the expectations of most people, the Government announced in 1998 and 1999 that it would legislate for a right to roam over open country. While hailed as a significant victory in the campaign for greater freedom of access to the countryside, Wilmore (1998) has already suggested that the definition of 'open country' will be hard to substantiate in law and will not, in any event, extend to any lowland areas of the country. The impact of such legislation is therefore unlikely to be significant in the context of this paper.

References

Baudrillard, J. (1981) *For a Critique of the Political Economy of the Sign* (trans. C. Levin). St. Louis: Telos Press.

Baudrillard, J. (1988) *Selected Writings* (ed. M. Poster). Stanford, Conn.: Stanford University Press.

Baudrillard, J. (1994) *Simulacra and Simulation* (trans. S.F. Glaser). Ann Arbor: University of Michigan Press.

Bishop, K.D. and Phillips, A.A.C. (1993) 'Seven steps to market – the development of the market-led approach to countryside conservation and recreation'. *Journal of Rural Studies* 9: 315–38.

Blunden, J. and Curry, N. (eds) (1989) *A People's Charter?* London: HMSO.

Bonyhady, T. (1987) *The Law of the Countryside. The Rights of the Public*. Abingdon: Professional Books.

Bosanquet, A. (1995) 'Minutes of evidence', in House of Commons Environment Committee, *The Environmental Impact of Leisure Activities, Volume II: Minutes of Evidence*, House of Commons Papers 246-II. London: HMSO, para 869.

Clark, G., Darrall, J., Grove-White, R., Macnaghten, P. and Urry, J. (1994) *Leisure Landscapes. Leisure, Culture and the English Countryside: Challenges and Conflicts*. Centre for the Study of Environmental Change, Lancaster University.

Clark, G., Darrall, J., Grove-White, R., Macnaghten, P. and Urry, J. (1994a) 'New approaches concerning public attitudes to the countryside: report on a research study'. Background Paper 8 in Clark *et al.*, *Leisure Landscapes*.

Couldry, N. (1998) 'The view from inside the simulacrum: visitors' tales from the set of Coronation Street'. *Leisure Studies* 17(2): 94–107.

Countryside Commission (1991) *Access Payment Schemes*. Discussion Paper. Cheltenham: Countryside Commission.

Countryside Commission (1991a) *Countryside Stewardship: An Outline*. Publication CCP 346. Cheltenham: Countryside Commission.

Countryside Commission (1992) *Enjoying the Countryside: Policies for People*. Publication CCP 371. Cheltenham: Countryside Commission.

Countryside Commission (1993) *National Targets for Rights of Way: The Milestones Approach*. Publication CCP 436. Cheltenham: Countryside Commission.

Countryside Commission (1994) *Countryside Stewardship: Handbook and Application Forms*. Publication CCP 453. Cheltenham: Countryside Commission.

Countryside Commission (1995) *Quality of Countryside: Quality of Life*. Publication CCP 470. Cheltenham: Countryside Commission.

Countryside Commission (1995a) *The Environmental Impact of Leisure Activities on the English Countryside*. Submission to the House of Commons Environment Committee. Cheltenham: Countryside Commission.

Countryside Commission (1996) *The Condition of England's Rights of Way: A Summary of the 1994 National Rights of Way Survey*. Publication CCP 505. Cheltenham: Countryside Commission.

Crouch, D. and Ravenscroft, N. (1995) 'Culture, social difference and the leisure experience: the example of consuming countryside', in McFee, G., Murphy, W. and

Whannel, G. (eds) *Leisure Cultures: Values, Genders, Lifestyles*. Publication No. 54. Eastbourne: Leisure Studies Association, pp. 289–302.

Curry, N.R. (1994) *Countryside Recreation, Access and Land Use Planning*. London: E. & F.N. Spon.

Curry, N.R. and Ravenscroft, N. (1997) 'The development of charging policies for the making of public path orders', *Local Government Studies* 23(2): 127–34.

Davies, G.H. (1995) 'Pay-off for conservation'. *Countryside* 76: 4–5.

de Certeau, M. (1984) *The Practice of Everyday Life*. Berkeley: University of California Press.

Department of the Environment and Ministry of Agriculture, Fisheries and Food (1995) *Rural England: A Nation Committed to a Living Countryside*, Cm 3016. London: HMSO.

Department of the Environment, Transport and the Regions (1998) *Access to the Open Countryside*. Consultation Paper. London: Department of the Environment, Transport and the Regions.

Department of the Environment, Transport and the Regions (1998a) *Meacher Makes Access to the Countryside a Top Priority*. News Release 129/ENV. London: Department of the Environment, Transport and the Regions.

Donzelot, J. (1980) *The Policing of Families*. London: Hutchinson.

Fitzpatrick, P. (1992) *The Mythology of Modern Law*. London: Routledge.

Foucault, M. (1980) *Power/Knowledge: Selected Interviews and Other Writings 1972–1977*. New York: Pantheon Books.

Glyptis, S. (1995) 'Recreation and the environment: challenging and changing relationships', in D. Leslie (ed.) *Tourism and Leisure: Towards the Millennium, Vol. 1: Culture, Heritage and Participation*. Publication No. 51. Eastbourne, UK: Leisure Studies Association, pp. 171–8.

Halfacree, K. (1993) 'Locality and social representation: space, discourse and alternative definitions of the rural'. *Journal of Rural Studies* 9(1): 23–37.

Hetherington, P. (1998) 'These boots are made for walking'. *Guardian* 24 Feb. 1998, p. 13.

Kay, G. and Moxham, N. (1996) 'Paths for whom? Countryside access for recreational walking'. *Leisure Studies* 15: 171–83.

Keith, M. and Pile, S. (1993) 'Introduction to part 2: the place of politics', in M. Keith and S. Pile (eds) *Place and the Politics of Identity*. London: Routledge, 22–40.

Labour Party (1997) *Manifesto*. London: Labour Party.

Lefebvre, H. (1991) *The Production of Space*. Oxford: Blackwell.

Marsden, T., Murdoch, J., Lowe, P., Munton, R. and Flynn, A. (1993) *Constructing the Countryside*. London: UCL Press.

Massey, D. (1993) 'Politics and space/time', in M. Keith and S. Pile (eds) *Place and the Politics of Identity*. London: Routledge, pp. 141–61.

Ministry of Agriculture, Fisheries and Food (1993) *Agriculture and England's Environment*. London: Ministry of Agriculture, Fisheries and Food.

Myerson, G. and Rydin, Y. (1996) *The Language of Environment. A New Rhetoric*. London: UCL Press.

National Trust Access Review Working Party (1995) *Open Countryside*. London: National Trust.

Newby, H. (1986) 'Towards a new understanding of access and accessibility (1). Values, attitudes and ideology', in Talbot-Ponsonby, H. (ed.) *Proceedings of the 1986 Countryside Recreation Conference*. Bristol: Countryside Recreation Research Advisory Group, pp. 14–24.

Parker, G. (1996) 'ELMs disease: stewardship, corporatism and citizenship in the English countryside'. *Journal of Rural Studies* 12(4): 399–411.

Ramblers' Association (1984) *Keep Out. The struggle for Public Access to the Hills and Mountains of Britain, 1884–1984*. London: Ramblers' Association.

Ramblers' Association (1993) *CSS – Survey by the Ramblers' Association of Public Access Sites*. London: Ramblers' Association.

Ramblers' Association (1994) *CSS Public Access Sites. Survey Report on Access Provisions at Second-Year Public Access Sites*. London: Ramblers' Association.

Ravenscroft, N. (1992) *Recreation Planning and Development*. Basingstoke: Macmillan.

Ravenscroft, N. (1995) 'Recreational access to the countryside of England and Wales: popular leisure as the legitimation of private property'. *Journal of Property Research* 12: 63–74.

Ravenscroft, N. (1998) 'Rights, citizenship and access to the countryside'. *Space and Polity* 2(1): 33–49.

Ravenscroft, N. and Reeves, J. (1997) 'Rural planning after the White Paper: new paradigm or old rhetoric?', in *Roots 97. Proceedings of the 1997 Rural Practice Research Conference of the Royal Institution of Chartered Surveyors*. London: Royal Institution of Chartered Surveyors, pp. 31–40.

Ravenscroft, N., Prag, P.A.B., Gibbard, R. and Markwell, S.S. (1995) *The Financial Implications to Landowners and Farmers of the Countryside Stewardship Scheme*. Research Series 95/1. Centre for Environment and Land Tenure Studies, University of Reading.

Ravenscroft, N., Markwell, S.S. and Curry, N.R. (1996) 'Evaluation of Highways Authorities' Milestones: Statements for Public Rights of Way'. Unpublished report to the Countryside Commission. Centre for Environment and Land Tenure Studies, University of Reading.

Ravenscroft, N., Gibbard, R., Markwell, S.S. and Reeves, J. (1998) 'Private sector agricultural tenancy arrangements: the European experience'. Paper presented to the Centre for Property Law conference 'Contemporary Issues in Property Law', University of Reading, 25–27 March 1998.

Rowan-Robinson, J. (1998) 'Rights of access'. Paper presented at the Centre for Property Law conference 'Contemporary Issues in Property Law', University of Reading, 25–27 March 1998.

Royal Institution of Chartered Surveyors (1989) *Managing the Countryside. Access, Recreation and Tourism*. London: Royal Institution of Chartered Surveyors.

Shoard, M. (1987) *This Land is Our Land. The Struggle for Britain's Countryside*. London: Paladin Grafton Books.

Spink, J. (1994) *Leisure and the Environment*. Oxford: Butterworth-Heinemann.

Thompson, E.P. (1993) *Customs in Common*. Harmondsworth: Penguin.

Urry, J. (1990) *The Tourist Gaze. Leisure and Travel in Contemporary Societies*. London: Sage.

Urry, J. (1995) *Consuming Places*. London: Routledge.

Warren, S. (1993) '"This heaven gives me migraines". The problems and promise of land-scapes of leisure', in J. Duncan and D. Ley (eds) *Place / Culture / Representation*. London: Routledge.

Werlen, B. (1993) *Society, Action and Space: An Alternative Human Geography* (trans. G. Wells). London: Routledge.

Whitby, M. and Lowe, P. (1994) 'The political and economic roots of environmental policy in agriculture', in M. Whitby (ed.) *Incentives for Countryside Management. The Case of Environmentally Sensitive Areas*. Wallingford, Oxon: CAB International, pp. 1–24.

Wilmore, C. (1998) 'Rights of access in England and Wales'. Paper presented at the Centre for Property Law conference 'Contemporary Issues in Property Law', University of Reading, 25–27 March 1998.

7

NARRATIVISED SPACES

The functions of story in the theme park

Deborah Philips

Introduction

The Disneyland park map is designed as an island – with blue and limitless boundaries that blend into an outside world that is at a remote distance. This has become the paradigm for the theme park map in which the theme park site becomes a utopic island, a world in itself. In its mapping, landscaping and in its naming, the theme park offers itself as a boundless world: Disneyland, Disney World. In Britain, Chessington offers a World of Adventures, while Thorpe Park invites the visitor to: 'Cross . . . into another world of fun and fantasy, where everything and everyone is larger than life. It's big, it's giant, it's humungous.' (Great Thorpe Park map 1998). The 'theming' of a theme park is what renders this strangeness domesticated. It is the employment of well-loved and recognised tales that makes the 'empty space' and alien territory of the theme park plea-surable and familiar. The knowledges that are brought to the theme park extend beyond the geographical: they involve sets of cultural knowledges drawn from both popular culture and from 'high' art.

If Marin has described the Disneyland sites as 'utopic space' (Marin 1984), theme parks are not utopian in the sense of Thomas More's island named for Nowhere; they do not declare themselves as complete alternative worlds (with the exception of Epcot, which as an 'Experimental Prototypical Community of Tomorrow', does make explicit claims towards the utopian, as does 'Celebration', Disney's new town in Florida). Rather, the theme park is a heterotopia, a compen-satory space, defined through its difference from the sites of work and of the everyday. In an interview, Foucault identified 'heterotopias' as: 'those singular spaces to be found in some given social spaces whose functions are different or even the opposite of others' (ed. Rabinow 1986: 252). In a later article on archi-tectural space, Foucault developed the concept, in terms which clearly include the theme park as a heterotopic space. He goes on to argue that the heterotopia both confirms and contradicts the sets of oppositions by which social space is organised, and among these is the distinction between work and leisure:

> Through the sets of relationships that define them, one could describe arrangements where one makes a temporary halt: cafés, cinemas, beaches . . . those which are endowed with the curious property of being in relation with all the others, but in such a way as to suspend, neutralize or invert the set of relationships designed, reflected or mirrored by themselves. These spaces, which are in rapport in some way with all the others, and yet contradict them . . .
>
> (Foucault 1986: 252)

The siting of a theme park space suggests this paradox of rapport and contradiction with the normative. A theme park is often situated beyond the city and suburbs, yet is easily accessible by car, and on familiar commuter routes. The park will often be built on a reclaimed site, an industrial or agricultural space that has been converted to leisure use. Disneyland was built on the site of former orange groves, Disney World in the Florida swamplands, Disneyland Paris on marshy land outside Paris. In England, Thorpe Park is sited on reclaimed gravel pits, and has refashioned a once working farm into an attraction, while both Chessington and Alton Towers are on the site of what were once country house estates.[1] The theme park is in a recognisable geographical location, but it makes no reference to the local landscape or culture. In Foucault's terms, it is: 'a sort of place that lies outside all places and yet is actually localizable' (Foucault 1986: 12). In Britain, the country houses that were once at the economic centre of the Chessington and Alton Towers parklands have become dislocated from their function and history, to become 'Burnt Stub Mansion' and 'The Towers', their architecture indistinguishable from the Gothic or Fairytale pastiche of the newly built attractions. For Giddens, such an erasure of local specificity can be described as 'disembedded space': 'By disembedding I mean the "lifting out" of social relations from local contexts of interaction and their restructuring across indefinite spans of time-space' (Giddens 1990: 21).

This 'lifting out' from local contexts was to become explicit Disney policy, after the experience of the incursion of local development around the first Disneyland site in Los Angeles. The later Walt Disney World in Florida was built on a site of 27,400 acres, precisely to protect the theme parks from the local environment, as an official Disney account explains:

> When it opened, Disneyland had been an island of fantasy set in an ocean of orange trees. That soon changed as urban sprawl arrived to surround the park with motels, gas stations, and fast-food joints. Walt

Disney watched these haphazard developments with displeasure and vowed that in Florida he would build a Magic Kingdom that was completely insulated from the outside world.

(Finch 1995: 405)

The theme park can attempt to completely insulate its site from the 'outside world', and it can reclaim an 'empty space' in so far as it is situated on land not considered viable for building, agriculture or industry. However, as Foucault points out:

we do not live in a homogenous and empty space, but in a space that is saturated with qualities, and that may even be pervaded by a spectral aura. The space of our primary perception, of our dreams and of our passions, holds within itself almost intrinsic qualities: it is light, ethereal, transparent or dark, uneven, cluttered, . . . it is a space of height, of peaks, or on the contrary, of the depths, of mud; space that flows, like spring water, or fixed space.

(Foucault 1986: 11)

Foucault is here referring to a psychological 'inner space', but his description evokes an image of the theme park's attractions; it is precisely these fantasies, 'the space of our primary perception, of our dreams and of our passions', to which the theme park lays claim. Disney explicitly promises an experience 'better than any dream you can ever have' (Disneyland promotional leaflet 1997), Chessington offers the visitor 'the ultimate day out of this world', while Alton Towers is advertised as a site 'where the magic never ends' (Chessington and Alton Towers Theme Park maps 1998).

The theme park is a space unequivocally devoted to pleasure, a zone to which families and groups travel in order to experience a world that is defined as a site of leisure, and which is especially visited at holiday times. It is defined as this not only through its organisation as a different kind of space, but also in that it offers a different organisation of time. Giddens has referred to the importance of the clock and of the calendar in ordering and distinguishing work and leisure time: 'The clock expressed a uniform dimension of "empty" time, quantified in such a way as to permit the precise designation of "zones" of the day (e.g. the working day)' (Giddens 1990: 17).

Theme park time is differentiated from the working day in that it is a space without clocks; a time zone of leisure. For Foucault: 'The heterotopia enters fully into function when men find themselves in a sort of total breach of their traditional time' (Foucualt 1986: 15). The 'parenthesis of time', in Stam's phrase (Stam 1988: 127) is an integral feature of the carnival space; clocks are absent, and so are the regularities of everyday life, meal times are not regulated, food is available at any time, and many restaurants serve all day breakfast. At Blackpool Tower and at Disney's Paradise Island, dancing can begin in the morning and go on all day in an artificially lighted space. Richard Hoggart, writing in the 1950s, speaks of these occasions of dislocated time as 'specific acts of baroque living':

> Most working class pleasures tend to be mass-pleasures, overcrowded and sprawling. Everyone wants to have fun at the same time, since most buzzers blow within an hour of each other. Special occasions – a wedding, a trip to the pantomime, a visit to the fair, a charabanc outing – assume this, and assume also that a really special splendour and glitter must be displayed.
>
> (Hoggart 1958: 116)

If the factory buzzers which determined leisure time for Hoggart's 1950s working class have long gone, it is still the case that 'Everyone wants to have fun at the same time'. As the rash of television commercials for theme parks at half-term breaks and Bank Holidays suggest, school timetables firmly structure leisure time for parents and children. Holidays and public festivals are days on which the fun fair sets up on a local common, and also when families and groups set out to congregate at leisure centres and theme parks. These Bank Holidays and public festivals remain still the ecclesiastical feast days which Bakhtin cites as the space of the carnivalesque (Bakhtin 1984: 19). As Disney marketing has shown, the theme park has become a site of ritual celebration for families and groups on important occasions.

The theming of pleasure zones

Theme parks are sites that are devoted to pleasure, but as Venturi and Scott Brown learnt from Las Vegas (Venturi, Scott Brown and Izenour 1977) the sites in themselves have no desirable features, but are spaces that are decorated and themed; what gives these spaces a character, what hails people to them and what makes them appear to be sites of fantasy and pleasure is the employment of story. The actual site of the theme park may be an unknown part of the city, a space outside the familiar environs of everyday life, a dislocated and disembedded space, but it is replete with familiar literary references. With rare exceptions (the references to Perrault's tales at Disneyland Paris are among the few acknowledgements on the site that the park is in France rather than America), these stories are imposed upon the space, rather than intrinsic to it. As Giddens notes:

the sense of the familiar is one often mediated by time-space distan-
ciation. It does not derive from the particularities of localised place . . .
The reassurance of the familiar, so important to a sense of ontological
security, is coupled with the realisation that what is comfortable and
nearby is actually an expression of distant events and was 'placed into'
the local environment rather than forming an organic development
within it.

(Giddens 1990: 140)

It is apparent that Chessington and Alton Towers do not know how to employ
the particularities of their sites; instead, the houses that were once the rationale
for the sites are now woven into narratives that have no relation to the locale.
Alton Towers, Chessington World of Adventures and Thorpe Park all import the
Western genre, among other distant events, into an English landscape, while all
the Disneyland sites recreate European folk and fairy tale in Los Angeles, Florida
and Tokyo as well as in Paris. For Susan Davis, there is a distinction to be made
between the diversity that is possible in fairgrounds, and the homogeneity of the
theme park:

The theme park replaced the carnival and midway entrepreneurs with
centrally produced and managed attractions and, just as important to
advertisers and sponsors, control over image and style. Within the
industry, the verb 'to theme' refers to this totalizing effect.

(Davis 1996: 40)

A major strategy for theming is the employment of stories – and of specific sets
of stories. In his analysis of the 'theming of America', Gottdiener has attempted
to identify the themes that structure the narratives of a nation in the theming
of America, and suggested that there are ten recurrent structuring narratives;
the spaces of the carnival site can similarly be identified as organised through a
strictly limited set of narratives. Gottdiener argues (in a listing that corresponds
to Foucault's description of heterotopias) that:

the themed milieu with its pervasive use of media culture motifs that
define an entire built space increasingly characterizes not only cities but
also suburban areas, shopping places, airports, recreation spaces such as
baseball stadia, museums, restaurants, and amusement parks.

(Gottdiener 1997: 4)

If the theme park is a bounded space, it nonetheless claims an infinite
variety: the Disney World promotional video promises: 'An endless array of

95

entertainment' (Disney World, Florida promotional video 1990). But it is not the infinite variety that it appears; as the theme park geographically literally disguises its boundaries, it also denies that there are limits to the range of stories it can tell. The stories that are regularly invoked in theme park sites (but – importantly – not exclusive to them) can be identified as belonging to a limited set; there is a restricted typology of genres that recur throughout theme parks and fairgrounds, and which are dominant at the Disneyland sites. The original Disneyland is important because it set up a paradigm on which almost all commercial theme parks are now structured, it is organised into different 'lands' – each of which corresponds to a popular narrative genre. Gottdiener attributes this organisation of the Disneyland sites to the discourses of Disney itself:

> Each realm of the Disneyland environment represents domains developed by the Disney Corporation in films and television shows over the years. They are: Tomorrowland – a sophomoric representation of futurism; Frontierland – the scene of old Disney successes, such as Davey Crockett and a representation of the early days of American settlement; Adventureland – escapist fantasies ranging from Tom Sawyer rafting to safari jungle cruises; and Fantasyland – the virtual site of Disney cartoon fantasies and animated films.
>
> (Gottdiener 1997: 111)

The theming of the Disneyland sites can however be identified more specifically than this; each of them corresponds to a popular literary genre: Tomorrowland (or Discoveryland in Disneyland Paris), is the site of Science Fiction narratives, Fantasyland, of Folk and Fairy Tale and the Gothic, while Adventureland invokes tales of Explorers and Treasure Islands, and Frontierland and Main Street USA, the Western. These are the generic categories that represent the dominant structuring of stories told in public sites of pleasure, it is these genres which have come to signify 'pleasure' and 'play'. While the theme park and fairground appear to offer an infinite possibility for narratives, they are in fact strictly limited to a particular set of genres, which can be endlessly combined: the television programme *X Files* (an attraction at Thorpe Park) is an example of the combination of tropes of the Gothic with Science Fiction, while *Star Wars* films, the basis of many an attraction, combine Science Fiction and Fairy Tale.

The genres which recurrently feature at carnival sites, whether Disney, fairground or theme park, all offer versions of a utopian sensibility; the narratives of contemporary carnival are less a celebration of past cultural glories (although they may often be that) than they are a search for other possibilities. The theme park is a part of the entertainment industry, and as Richard Dyer has explained:

Entertainment offers the image of 'something better' to escape into, or something we want deeply that our day-to day lives don't provide. Alternatives, hopes, wishes, these are the stuff of utopia, the sense that things could be better, that something other than what is can be imagined and maybe realised.

(Dyer 1992: 18)

Each of the genres employed at the theme park offers the possibility that things could be other, that the world could be reimagined: at the moment of Camelot (both Arthurian, and in an American context, of the Kennedy years), of the American pioneer, of the space launch. These are stories which are repeatedly told in different contexts, and which involve the fantasy of the colonisation of another world. The structural regularities of the genres and their associated fantasies of colonisation in the theme park can be identified as:

Fairy and Folk Tale (a dream of harmony with nature and other worldliness)
Gothic (the conquering of the supernatural)
Chivalric Romance (the 'Golden Age of Camelot')
Explorers and Treasure Islands (the colonisation of geographies)
Science Fiction (the conquering of new worlds in space)
Westerns (the pioneering spirit conquering new frontiers).

According to Maria Corti: 'Literature is subject to structuring on various planes or levels' (Corti 1978: 2): one of the planes on which the 'literary' is structured is the site of the carnivalesque, a space in which signs connoting a set of literary texts can be assumed to bind an audience. Narratives are often signified in the carnival space through the metonymic representation of a defining character: *The Wind in the Willows* is referenced through the image of a frog dressed in waistcoat and spats, to become *Toad of Toad Hall*. A whole text is invoked through synecdoche, which may refer to a single object in the narrative. An attraction organised around revolving teacups (to be found in all the Disney sites and at Alton Towers and Thorpe Park) references the mad hatter's tea party, and so invokes *Alice in Wonderland*.

Narratives of pleasure

The genres that are evoked in the theme park are those that are familiar from childhood reading; there are constant references in the theming of attractions to stories that are part of a global cultural vocabulary. The references at Disney are not only to film and television, but also to a range of literary icons: H.G. Wells, Jules Verne, Mark Twain and Charles Perrault are among the authors recurrently and directly invoked in the theming of Disneyland attractions and in their promotional material. *Alice in Wonderland*, *The Wind in the Willows*, *Snow White*, *Winnie the Pooh* and *The Sword in the Stone* are all texts which provide the basis for major attractions at Disneyland. If Rider Haggard and Conan Doyle, Hans Christian Andersen, the Brothers Grimm, Bram Stoker, Malory and Tennyson are not directly acknowledged, variations of their stories are to be found in Disneyland, and in almost all other Western theme parks.

The majority of stories to be found in theme parks are narratives that are associated with childhood: fairy and anthropomorphic tales. It is the familiarity of these stories that renders the unfamiliar space of the theme park desirable and comfortable. These are stories that may be or may not be known through the original text, but will be known through their circulation across a range of media. Evoked not only in print, but in the imagery of marketing and of theming, they are narratives that have a multi-generational appeal and recognition. These are stories that are mediated through abridged versions for children, illustrated children's anthologies, cartoons, animation and film to become stories that enter into a global cultural vocabulary.

The range of stories in the theme park and Disneyland does apparently offer a polyphony of narratives, an apparently infinite variety of stories and a clash of genres.[2] The range of narratives offered in the theme park is, however, strictly generically limited – and they are not told by a variety of independent voices. The narrator at Disneyland is always Disney; there is only one valid voice. Other theme parks emulate this construction of a single narrator; at Chessington, there is the awkward fictional creation 'Dr Chessington' whose invention the theme park is. Mr Rabbit leads the Thorpe Park Rangers at Thorpe Park, in a hierarchy of elephants and jungle animals who tell the tales. 'A giant bookworm' reads the stories of Alton Towers. Whether the narrator be Walt Disney, Dr Chessington or a worm, these are all fictional constructions of a corporate identity; Mickey Mouse may insert himself as the hero of a wide variety of stories, but he is also a corporate logo. The spaces and attractions of the theme park do answer in some measure to popular memory, but they also impose a very particular framework from which to read its stories.

Disney has become central to this process and a direct conduit for the importing of literary and folk tales into sites of pleasure and into the mass market realm of popular culture. The sequence of transposition from tale to film to theme park attraction and into themed commodities for children, has led to a dominance of those particular images of folk tale heroes and heroines. The hold that the Disney corporation has through its marketing skills across the whole range of the media, means that Disney iconography, what Schickel has called 'The Disney Version' (Schickel 1968) predominates and informs leisure sites across the globe. Sharon Zukin and Louis Marin (Zukin 1991, Marin 1984) are among the many other critics who have seen Disney as a totalitarian onslaught on the world's favourite stories.

If these stories are commodified in the theme park, however, their telling cannot be entirely contained by it. The tales that are told cannot be entirely subsumed, for that is not the only place in which they are told. The stories that recur in these sites, have what Bakhtin has called 'prior speakings',[3] they are tales that have already been told, and will continue to be told in different versions. The tales and the elements of the carnivalesque on which the theme park depends for its construction of 'pleasurable' experience are dynamic structures, however commodified their pleasures might be.

Narratives are among our earliest and most significant childhood experiences; before we are able to read ourselves we are told stories, and are read to by parents or by teachers. The evocation of a narrative invokes not only the story itself, but also the memory of its telling. Stories can bring back powerful memories of a secure and familiar childhood experience, a form of gratification in which a parent or guardian gave the child undivided attention. These stories are a source of pleasure not only to the children to whom they are ostensibly addressed, but also to adults, for whom these stories evoke a nostalgia for their own childhood. Recent figures for Disneyland visitors suggest that adult visitors outnumber children four to one, indicating that some visitors are not accompanied by children at all. In 1994, the Disneyland Paris park was found in a marketing survey to have 'had a significant appeal to the older generation (aged 55 and over) who came without children' (Bryman 1995: 88–9). The prevalence of 1950s icons and styles in theme parks is indicative of an appeal to a baby boom generation, who are deriving pleasure from revisiting their own relationship to popular culture, and to the stories that they themselves encountered in childhood. As Bryman points out:

> children could discover the new anew while adults enjoy the memory
> of that discovery as children (seeing Peter Pan, Swiss Family Robinson,
> Cinderella and other Disney films which have spawned attractions), as
> well as nostalgic recollections of the kind of place symbolized by Main
> Street.
>
> (Bryman 1995: 89)

The function of stories, as Bettelheim has explored in relation to fairy tale (Bettelheim 1978), is also to provide a point of entry into the mysteries of adult life. Stories are a means not only for children but also for adults to play with the different possibilities of narrative resolution; they are a means of addressing and articulating unease, and allow for a way of imagining that things could be different. The stories which recur in pleasure spaces offer a magical resolution (a happy ever after) to fictional and real frustrations. The stories that recur in pleasure sites provide a space for the articulation of aspirations and of nostalgias for a different kind of world and community. Stories are, as Fredric Jameson has argued, integral to 'the Political Unconscious' (Jameson 1981) – they are articulations of fantasies, desires and repression, and like repressed wishes, they constantly recur. It is narratives which render the unfamiliar world of the leisure space familiar and desirable and which evoke a pleasurable nostalgia.

Childhood stories are not only about nostalgia, but also about hopes and dreams that have been unrealised. The theme park, and especially Disneyland, manifests a preoccupation with the past, with childhood, and for a kinder, gentler time that has gone. Disneyland promises (literally) a return to that childhood experience: the promotional video offers the chance of: 'seeing the world as it should be – through the eyes of the child' (promotional video 1996).

The genres of the theme park

The theme park consistently promises 'magic' and enchantment; Fairy Tale is the genre most associated with childhood, and is given its own space in the theme park: Fantasyland at Disney sites, the Land of Make Believe at Alton Towers. Snow White, Cinderella, Sleeping Beauty are all central narratives to the theme park, stories which are derived from folk tale, but which have been variously written by Perrault, Anderson and Grimm (it is important to recognise that Disney is only one in a long line of tellers of these tales). In each Disneyland, a fairy tale castle is the focal point of the park, a distant apparition waiting to be explored. This is, in fact, the final frame of the Disney fairy tale film, the castle to which the Prince transports his princess. Of the genres offered in the theme park, it is only Fairy Tale that directly addresses the feminine. Fairy Tale is one of the few spaces in the theme park and fairground that signals its appropriateness to young children, and particularly to girls. Romance, the one popular fictional genre written entirely by and for women, is mapped on to the fairy tale in the theme park, and is one of the few sites at which the genre is to be found.

Fantasyland also features the Gothic: with its associations of the supernatural and death, it might be expected that the Gothic would be absent at Disneyland, but it recurs in several sites; the Snow White train has all the characteristics of a Ghost Train, including skeletons, bats, cobwebs and tombs. There are also references to an American Gothic tradition; the Haunted House attraction is modelled on an Edgar Allan Poe story. Every fairground and theme park has a ghost train or haunted house which replicates combinations of images drawn from nineteenth century Gothic fiction, of which Bram Stoker's *Dracula* and Mary Shelley's *Frankenstein* are the most frequently invoked. Chessington features a 'Transylvania'

100

site, which contains a mad professor's laboratory, and a 'Vampyre' ride, while Alton Towers has a 'Gloomy Wood' and Haunted House.

Anthropomorphic narratives in the theme park are often drawn from folk tale, but also derive from such children's classics (often filmed by Disney) as: *Winnie the Pooh, The Wind in the Willows, The Jungle Book*. The use of animals is one of the most evident instances of a Bakhtinian carnival in the theme park site; anthropomorphism allows for a carnivalesque celebration of the body, in that it displaces the grotesque body onto the animal, and at one remove from the human. 'Family' theme parks would shy away from any overt celebration of excess or of the sexual, but the representation of animal stories allows for a displacement of unacceptable qualities of gross appetites, rage and desire. Anthropomorphised animals are a recurrent feature of the theme park: Alton Towers employs variations of Beatrix Potter's *Peter Rabbit* and *Squirrel Nutkin,* Thorpe Park a cast of animal 'rangers' who include Miss Hippo, Mr Rabbit and Mr Giraffe.

There is a consistent use of chivalric and heraldic iconography in fairgrounds and theme parks, which construct a nostalgic version of a history when men were heroes and women were there to be rescued. The employment of Arthurian references in the theme park is one that derives from nineteenth-century versions of the Chivalric, and offers a paradigm for understanding the process of the transformation of 'high' cultural narratives into popular forms. Images of Arthurian Romance are to be repeatedly found in carousels, flags and in replicas of castles. Thorpe Park has a 'Courtyard Challenge' complete with jousting regalia, the central carousel at Disneyland Paris features Mickey Mouse as 'Sir Lancelot'. The removal of the sword in the stone (overseen by Merlin) is a daily ritual at the Disney World and Disneyland sites. Such signifiers of medievalism are the legacy of a set of mediations that can be traced from Malory and Tennyson's poetic reworkings of Arthurian legend, through to T.H. White's Camelot novel *The Sword in the Stone*, later filmed by Disney, and to the musical and film of *Camelot*.

Adventureland is the site at Disney of two recurring popular genres: the Travel and Explorer narrative. Adventure stories have become the basis of the staple attractions of any theme park; Chessington's World of Adventures features a 'Forbidden Kingdom' and 'Mystic East'; Alton Towers, a 'Forbidden Valley' and 'Congo River Rapids'. Thorpe Park's animal rangers are jungle explorers too; Miss Hippo is situated in a 'Fungle Safari', while Mr Rabbit invites the visitor to share his 'Tropical Travels'. These are stories which are reconfigurations of *Boy's Own* 'Stories of the Empire', and of such novels as Rider Haggard's *King Solomon's Mines*. The pith-helmeted explorer is a colonial figure who constantly reappears in various guises, most familiarly as Indiana Jones. The narrative of the colonisation of unknown landscapes is perhaps the most recurrent at the theme park, and is also to be found in tales of piracy and lost islands.

The explorer and the pirate are both the protagonists of a form of travel narrative, which is also to be found in science fiction. Macherey has described the ideological importance in what he calls the 'geographical novel' of mapping in the construction of geographical knowledges:

> By means of a map the journey is a conquest of the same sort as a scientific adventure. It recreates nature, in so far as it imposes its own norms upon it. The inventory is a form of organisation, and thus of invention. This is the meaning of the geographical novel.
>
> (Macherey 1978: 183)

Robert Louis Stevenson's *Treasure Island* is among the most frequently evoked stories in the theming and naming of carnival spaces, and one which frequently employs the map in its theming. The pirate, invoked in many attractions based around ships or water, most famously at Disneyland's 'Pirates of the Caribbean', is a figure who allows for an expression of masculine rebellion while simultaneously celebrating enterprise. Echoes of *Robinson Crusoe* are regularly to be found in the recurrent theme park fantasy of the colonising of an exotic island. The Swiss Family Robinson Treehouse at the Disneyland sites, modelled on the 1960 Disney film and derived from Johann Wyss's novel, is only one of the more explicit of these references.

The two genres which most explicitly articulate nostalgia in the theme park are those most directly concerned with the defining of frontiers: Science Fiction and the Western. These are the genres that structure the literal geography of Disneyland: as Marin has pointed out, the Monorail (coded as futuristic) and the Disneyland express train (coded nineteenth-century Western) are the only two attractions which span the entire site.

The Western is an indigenous American genre, but it has also signified 'modernity' in European carnival sites since the nineteenth century; the first privately owned pleasure site at Blackpool was named 'Uncle Tom's Cabin' and featured cowboy theming (Turner and Palmer 1981). The Western takes up two sites at Disneyland, Frontierland and Main Street. Along with the spirit of the American pioneer, the Western narrative is employed to celebrate a nostalgic fantasy of a pre-industrial community. The most elaborate reconstruction of this is to be found in Main Street, USA, a space which is to be found at all the Disneylands. Marin has identified Main Street as belonging to the Western genre, and its *mise en scène* is recognisable from Hollywood and television versions of Owen Wister's *The Virginian* and from Zane Grey's fiction. Main Street represents the most visible example of nostalgia at Disneyland for a knowable community. It is perhaps the theme park space which most directly embodies a simulacrum, in that it reconstructs a late nineteenth-century American small town street which could never have existed; all the 'Victorian' shops are supplied with goods from multi-national corporations.

Space represents yet another frontier, and involves an imagining of the future; its Disneyland sites are named Tomorrowland, Futureland, Discoveryland, but the attractions which should be the least likely articulations of nostalgia are those which manifest the most. Tales of the future, which might be a celebration of contemporary technology, are in those theme park sites that are firmly rooted in the past. The iconography of Disney's most recent attraction at Disneyland Paris, Space Mountain, is quite self-consciously drawn from the rocket design in an early film by Méliès. 'Science' and the 'Future' belong in theme parks to the nineteenth century and to the 1950s, the last moments at which technology could be unproblematically celebrated. The stories of Science and of Progress to be found in the theme park do not extend to the contemporary, but instead are derived from the nineteenth-century tales of Jules Verne and H.G. Wells and from 1950s comic book heroes, such as Dan Dare.

The ambivalence towards the Modern displayed in the 'futuristic' attractions which might be expected to embrace it is also apparent from the dominance of stories across the theme park which are drawn from the nineteenth century. The images which reference these narratives are overwhelmingly derived from nine-teenth-century popularisations of folk and fairy tale and legend (Tennyson, Andersen and Grimm, for example) or from 'classic' nineteenth-century fictions (Stevenson, Rider Haggard, Stoker).

Narratives of modernity

The collage of narratives to be found in the theme park is one of the features that Foucault cites as characteristic of the heterotopia:

> The heterotopia has the power of juxtaposing in a single real place different spaces and locations that are incompatible with each other. . . .
> Perhaps the oldest example of these heterotopias in the form of contra-dictory locations is the garden.
>
> (Foucault 1986: 14)

This coexistence of 'contradictory locations' and stories has led to a preoccu-pation with the theme park (and especially Disney) from postmodern critics, notably Baudrillard, who has argued that 'Disneyland is presented as imaginary in order to make us believe that the rest is real . . . ' (ed. Poster 1988: 172). Fjellman is among the many postmodern analysts of Disney to argue that at Disney World:

we have the disconnected, discrete space-time packages of postmodernism – bounded, localised themes placed next to each other serially and, for the most part, randomly.

(Fjellman 1992: 61)

For Giddens, such a collapsing of time and space is by no means symptomatic of postmodernity, but rather a consequence of modernity:

The dynamism of modernity derives from the *separation of time and space* and their recombination in forms which permit the precise time-space 'zoning' of social life; the *disembedding* of social systems (a phenomenon which connects closely with the factors involved in time-space separation); and the *reflexive ordering and reordering* of social relations in the light of continual inputs of knowledge affecting the actions of individuals and groups.

(Giddens 1990: 16–17)

The theme park, and especially Disneyland, can be read less as a product of a post-modern aesthetic than as the consequence of events designed to celebrate modernity: Disneyland, and Epcot in particular, has a direct lineage back to world fairs and to trade exhibitions and both redeploy attractions that were first developed by Disney for the New York World's Fair. The world fairs were designed to show off the commodities and scientific prowess of trading nations, and to celebrate the Progress that made it possible to bring these products together. In writing of the 1896 Berlin Trade Exhibition, Simmel speaks of Berlin becoming: 'a single city to which the whole world sends its products and where all the important styles of the present cultural world are put on display' (Simmel 1991: 120).

The World Showcase at Epcot can be read as a direct descendant of the world fair mission to assemble the products of the trading world in a single place. Simmel refers to the 'outward unity' of the Trade Fair which seems to integrate a myriad of cultural, social and commercial phenomena:

It is a particular attraction of world fairs that they form a momentary centre of world civilization, assembling the products of the entire world in a confined space as if in a single picture. Put the other way round, a single city has broadened into the totality of cultural production. No important product is missing, and though much of the material and samples have been brought together from the whole world they have attained a conclusive form and become part of a single whole.

(Simmel 1991: 120)

This 'conclusive form' is exactly what the theme park emulates, and is at its most evident in the Epcot World Showcase. The 'unity' of Disney is not a 'floating psychological idea'; in the late twentieth century that unity can become a company logo and a familiar brand name that has a firm material and global basis. The theme park assembles the products and the stories of the world not into a single city, but in a bounded site outside the city. The imposition of non-indigenous products applies to every aspect of the theme park, and is at its most acute at the Disneyland sites. A Walt Disney produced book proudly boasts the range of its imported detail:

> The landscape at Disneyland is 90 percent non-indigenous to California. Many of its more than 800 species of plants and trees are actually native to Australia, Mexico, Europe and Asia. Except for a botanical garden, Disneyland in fact features the largest collection of plant species found anywhere in the world.
>
> (Walt Disney Company 1996: 156)

For Foucault, such an accumulation of incompatible objects and styles in the heterotopia is less postmodern packaging than it is a consequence of a nineteenth century modernity:

> The idea of accumulating everything, . . . of creating a sort of universal archive, the desire to enclose all times, all eras, forms and styles within a single place, the concept of making all times into one place, and yet a place that is outside time, inaccessible to the wear and tear of the years, according to a plan of almost perpetual and unlimited accumulation within an irremovable place, all this belongs entirely to our modern outlook. Museums and libraries are heterotopias typical of 19th century Western culture.
>
> (Foucault 1986: 15)

If the theme park is not a museum or library (although both museums and libraries are increasingly using the marketing techniques and attractions of the theme park in order to compete for funding), it can be similarly read as a space which claims 'perpetual and unlimited accumulation'. This accumulation is at its most apparent in the piling of commodities from around the world, but there is a similar accumulation of stories. The tales that are told in the theme park are not (as Gottdiener argues) only those of American popular culture, but drawn from the cultures of the world.

The theme park offers a simulacrum of a collective place, nostalgia for an organic community, but it is an 'empty space' that has no investment in the surrounding locale or in any knowledge of its history or geography.

Giddens has differentiated between 'space' and 'place', in a distinction that helps to make sense of theme park geographies:

> 'Place' is best conceptualised by means of the idea of locale, which refers to the physical settings of social activity as situated geographically. . . . The advent of modernity increasingly tears space away from place by fostering relations between 'absent' others, locationally distant from any given situation of face-to-face interaction. In conditions of modernity, place becomes increasingly phantasmagoric: that is to say, locales are thoroughly penetrated by and shaped in terms of social influences quite distant from them. What structures the locale is not simply that which is present on the scene; the 'visible form' of the locale conceals the distanciated relations which determine its nature.
>
> (Giddens 1990: 19)

The theme park explicitly offers a 'phantasmagoria': it celebrates the fact that it can bring together 'absent others', and revels in the exoticism of its attractions. The theme park is a space which is unapologetically penetrated by influences quite distant from their geographical location, and which distances itself from its actual locale.

Our knowledge of the geographical site of the theme park itself and of the geographical spaces that it offers the visitor is already imbued with a range of familiar stories. The space of the theme park is saturated with geographical knowledges, but it also contributes to the making of lay knowledge. The stories that it tells frame particular and limited versions of art, history and geographical knowledges, and impose familiar and globalised versions over any local or community knowledge. The knowledges that are confirmed and reinscribed in the theme park continue to be those of the grand narratives of the nineteenth century: the stories that the theme park tells are those of the Empire and of colonial adventure, of Victorian heroes and heroines, and of a belief in Progress and the excitement of Science. The fantasy lands of the theme park claim to offer new experiences, new stories, new geographical knowledges, yet while claiming to expand our knowledge the theme park only serves to confirm the knowledges with which we arrived.

Notes

1 See Philips, Deborah and Haywood, Ian *Brave New Causes: postwar popular fictions for women* Cassell: London, 1998 for an account of the 'country house problem' in the postwar period.

2 See Philips, Deborah 'The commodification of the carnivalesque' in *Tourism and Tourist Attractions* ed. Bennett, Marion, Philips, Deborah and Ravenscroft, Neil, LSA Publications: Eastbourne, 1998 for an expansion of this point.

3 See Gardiner, Michael *The Dialogics of Critique: M.M. Bakhtin & the Theory of Ideology*, Routledge: London, 1992, pp. 23–31 for an account of this concept.

References

Bakhtin M. (1984) *Rabelais and His World* (trans. H. Iswolsky), Indiana University Press: Bloomington.

Bettelheim B. (1978) *The Uses of Enchantment: the meaning and importance of fairy tales*, Peregrine: Harmondsworth.

Bryman A. (1995) *Disney and His Worlds,* Routledge: London.

Corti M. (1978) *An Introduction to Literary Semiotics* (trans. M. Bogat and A. Mandelbaum), Indiana University Press: Bloomington.

Davis S. G. (1996) 'The theme park: global industry and cultural form', *Media Culture and Society* 18: 399–422.

Dyer R. (1992) *Only Entertainment*, Routledge: London.

Finch C. (for the Walt Disney Company) (1995) *The Art of Walt Disney: from Mickey Mouse to the Magic Kingdoms*, Virgin Books: London.

Fjellman S. M. (1992) *Vinyl Leaves: Walt Disney World and America*, Westview Press: Oxford.

Foucault, M. (1986) 'Other spaces: the principles of heterotopia', *Lotus International: Quarterly Architectural Review* 48/49: 9–17.

Gardiner M. (1992) *The Dialogics of Critique: M.M. Bakhtin and the theory of ideology*, Routledge: London.

Giddens A. (1990) *The Consequences of Modernity,* Polity Press: Cambridge.

Gottdiener M. (1997) *The Theming of America: dreams, visions and commercial Spaces*, Westview Press: Colorado.

Hoggart R. (1958) *The Uses of Literacy*, Penguin Books: Harmondsworth.

Jameson F. (1981) *The Political Unconscious: narrative as a socially symbolic act*, Cornell University Press: Ithaca.

Macherey P. (1978) *A Theory of Literary Production* (trans. G. Wall), Routledge and Kegan Paul: London.

Marin L. (1984) *Utopics: spatial play*, (trans. R. A. Vollrath) Macmillan: London.

Philips D. (1998) 'The commodification of the carnivalesque' in Bennett, M., Philips, D. and Ravenscroft, N. (eds) *Tourism and Tourist Attractions*, LSA Publications: Eastbourne.

Philips D. and Haywood I. (1998) *Brave New Causes: postwar popular fictions for women*, Cassell: London.

Poster M. (ed.) (1988) *Jean Baudrillard: selected writings*, Stanford University Press: Stanford.

Rabinow P. (ed.) (1986) *The Foucault Reader*, Peregrine Books: Harmondsworth.

Schickel R. (1968) *The Disney Version*, Simon and Schuster: New York.

Simmel G. (1991) 'The Berlin Trade Exhibition', *Theory, Culture and Society* 8(3): 119–25.

Stam, R. (1988) 'Mikhail Bakhtin and Left cultural critique', in Kaplan, E. A. (ed.) *Postmodernism and its Discontents: theories, practices*, Verso: London.

Turner B. and Palmer S. (1981) *The Blackpool Story*, The Corporation of Blackpool: Blackpool.

Venturi R., Scott Brown D. and Izenour S. (1977) *Learning from Las Vegas*, MIT Press: Cambridge.

Walt Disney Company (1996) *Walt Disney World*, Orlando, Florida.

Zukin S. (1991) *Landscapes of Power: from Detroit to Disneyworld*, University of California Press: Berkeley.

8

CULTURAL CONTESTATION AT DISNEYLAND PARIS

Stacy Warren

In the 1940s, the surrealist artist Salvador Dalí was hired by Walt Disney to illustrate a new generation of experimental cartoons. Dalí leapt into his job with exuberance at Disney's Burbank studios; the improbability of the historical moment was captured in a photograph taken in Disney's backyard, with Dalí primly seated upon Walt's 1/8-scale model train, hands crossed in his lap, a look of knowing satisfaction on his face (Marling 1997: 40–1). His Disney career lasted fifteen seconds. Dalí was promptly fired after the studios received the first animated frames of Technicolor melting watches, broken statuary, and figures meta-morphosing into turtles, ants, and dandelion puffs.

(Doss 1997: 186–7)

Relations between Europe and Disney have been peculiar ever since.

Introduction: into the dreamspace of desired modernity

Disney theme parks are the tourist meccas of the late twentieth century. Each year, more people visit Walt Disney World than the monuments of the United States capital; more visit Disneyland Paris than the Eiffel Tower. These facts are routinely heralded in Walt Disney Company press releases, making explicit the intended relationship between commodified entertainment and national symbol. Beneath these statistical tourist achievements, however, lie complex cultural inter-actions that undermine any sense of homogeneous experience or meaning. Each theme park beckons, promising to reveal to the careful observer the rich dynamic of its site and situation. The more cultural and ethnic layerings that comprise a park's everyday existence, the more intriguing the dynamic – and no Disney park has had a more convoluted existence than the one located in former beet fields 32 kms outside Paris.

This chapter investigates the cultural conflicts embedded in what has become the most popular tourist destination in France, Disneyland Paris. Disneyland Paris is, of course, a transplant of the extraordinarily popular American version of theme park. Yet almost continual friction between national identities (American and European; American and French; French and other European) has guaranteed the European theme park an exceedingly unique history that far transcends its role as a leisure centre. It was not an immediate success; in fact, sluggish attendance figures combined with outright hostility to Disney's presence culminated in a highly publicized financial crisis within two years. After extensive financial restructuring and cultural reshaping, as well as several name changes, the theme park began to attract more visitors and less negative publicity. In its new, more successful incarnation, Disneyland Paris has become an intriguing hybrid of conflicting French, European, and American values. As I will argue, Disneyland Paris has unintentionally become a site of postcolonial struggle in Disney's most ambitious recolonisation project to date.

To write about Disney is to visit an intellectual theme park where Disney criticism conducts its rides and spins its occasionally revelatory, more often creepy, wonders and illusions (Marcus 1997: 201). The first Disney theme park, Southern California's Disneyland, opened in 1955 and by the early 1960s the first academic critiques of the park began to appear. Four decades later there are now four Disney theme parks in three countries; Disney criticism is reaching a fever pitch. With each theme park, Disney realises a new extreme in the articulation of cultural dynamics. Now that Disney has crossed international borders, we are confronted with a new question: If Disneyland was about spectacle and Walt Disney World about the hyperreal, what will Disneyland Paris reveal?

Intellectual critics, however, much like European tourists, for the most part have ignored the French park. After an initial furore following French observer Ariane Mnouchkine's condemnation of the not-yet-opened EuroDisney (as Disneyland Paris was originally called) as a cultural Chernobyl, interest waned. Marcus (1997) lingers long enough to note that critics of the park rarely bother to 'ride the rides', for example, and Baudrillard (1996) takes a few fairly predictable swipes, but in general there has been very little direct and sustained engagement with it. The most penetrating Disney scholarship remains focused on the two American parks (Fjellman 1992, The Project on Disney 1995, Sorkin 1992, Zukin 1995). Yet even this, as Marcus (1997) points out, often slips into elitist condemnation without ever interrogating what happens in the Disney parks at ground level. In this chapter, I want to shift the gaze from American parks to the sole European one, and develop a theoretical framework that addresses Marcus' concern regarding Disney scholarship. I adopt a postcolonial sensibility, both because it offers a nuanced perspective from which to examine intercultural exchange, and because by making explicit the voice and position of the speaker, it confronts many of the weaknesses identified with conventional Disney scholarship. Consequently, this work proceeds in the spirit of Salvador Dalí cheerfully riding Walt's model train – a leap into the fray, as it were. Though I have never been to Disneyland Paris, I attempt to situate this work on the inside by incorporating the words and experiences of those who have, either as visitors or as employees. The chapter proceeds along the tourist's path and recreates, in effect, the intellectual theme park to which Marcus refers: the approach, arrival at the gates, and then visits to individual attractions. My goal is to explore, by opening up an overlooked area of Disney research, a broader constellation of issues concerning tourism, nation, commodified entertainment, and the cultural landscape. Our first stop, before even entering the park, is TheoryLand, where I introduce the theoretical framework that will guide the tour.

TheoryLand

Disneyland Paris inherited more than just a site plan from its American counterparts. California's Disneyland and Florida's Walt Disney World, by the early 1990s, had earned reputations as the coquettishly sinister manifestations of all that was wrong with American mass culture. A variety of intellectual observers, who can be loosely grouped together as 'postmodern', singled out Disneyland as supremely prophetic and paradigmatic of these cultural processes (Jameson, interviewed by Stephanson 1988: 8); Disneyland was the phantasmagoric playground where European postmodern thinkers discovered their worst theoretical constructs come true (Baudrillard 1983, Eco 1986, Marin 1984).

When Walt Disney World opened in 1971, critics pointed to the degree to which the lines between the real world and Disney's world blurred. Eco (1986:47) called the park no less than a model of an urban agglomerate of the future. While some commentators embraced Walt Disney World as urban space that worked (see especially Blake 1972), postmodern observers grew alarmed by the implications of the captive metropolis, noting in particular the increased propensity for Disneyesque fragmentation, simulacra, and commodification to be incorporated into real world urban spaces.

Disneyland Paris follows in this path of exponential transformation. When Disney came to Paris, it planted its postmodern footprint upon the land whose intellectuals invented it as postmodern space in the first place. It is Disney's most daring recolonization project. If Disneyland is about the colonization of the world (MacKay 1997), Disneyland Paris is something else again: it is about the playful recolonization of images of itself, embedded within a quietly sinister recolonization of economic, political, and cultural relationships between corporation, nation-state, and locality. A close examination of Disneyland Paris' history and everyday life reveals what Disney's postmodern critics have thus far missed: the theme park actually is a site of twisted postcolonial cultural debate. The cultural conflicts that occur within the park echo, within their safely manufactured confines, the conditions of postcolonial struggle in the so-called real world. In ways that the American parks or even Tokyo Disneyland will never approach, the intrusion of Disney into France has forced Disney employees, guests, and the general public to confront questions of 'us' and 'them', 'self' and 'other', 'intelligentsia' and 'subaltern', and other tropes of postcolonial discourse.

Postcolonial discourse, a mode of thought that recently has been popularized across the social sciences, traditionally has been applied to understanding the relationship between former colonial powers and the lands under their control. Motivated by the works of diasporic intellectuals such as Edward Said, Gayatri Spivak, and Homi Bhabha, it aims at both deeper understanding of these colonial histories and the Western scholarship that has (re)presented them (Frankenberg and Mani 1993). Defining colonialism as the conjunction between a politico-economic reality and a system of cultural representation, postcolonial literature interrogates the seemingly neutral, apolitical Western voices that naturalized colonial ideologies into universal, transhistorical 'truths' (Lee 1997: 89). Critical examination of traditional Western canons strips labels away and dissolves assumed identities: Western colonial nations are revealed as 'Eurocenter,' nations that shared economic, political, and military advantages that allowed them not only to control others but culturally to convince themselves and their subjects (at least some of their subjects) that they were the 'dreamspace of desired modernity' (Lavie and Swedenberg 1996: 158). The people subjugated under their rule were 'primitivized', assigned new identities that simultaneously exoticized and infantilized – and ultimately dehumanized – them in the eyes of the West (Emberely 1993, Lavie and Swedenberg 1996). Thus the condition 'Third World' is doubly a creation of the West: first, by establishing unequal political and economic relationships that ensured Western dominance, and second by defining so-called Third World populations as wholly 'other' and in every way inferior to Western people.

I argue in this chapter that the Walt Disney Company deserves honorary colonial status, especially given their dealings in France, and that the everyday tensions articulated at the park are generated at the intersection of neocolonial and postcolonial processes. This aspect of their existence has not been addressed by postmodern theorists or others writing on Disney's wider cultural significance. I propose that four fundamental concerns of postcolonial thought can provide us with points of departure to deepen our understanding of a place like Disneyland Paris. The first three address the nature of cultural communication as crafted by Disney:

1   postcolonial interest in the ways that colonial voices celebrate fabricated differences and speak not about but for others;

2   colonial control of local 'othered' populations; and

3   colonial economic exploitation.

The final postcolonial concern will help situate examination of the entire Disney phenomenon – beyond just the message Disney wishes to portray – in a theoretically sophisticated light:

4   the hegemonic destabilization that underlies colonial relationships.

After discussing the general historical scope of Disney's Recolonization Project from the first Disneyland to the present day, I examine each postcolonial concern in the context of its articulation at Disneyland Paris.

The Recolonization Project

> Meanwhile, from the bottom of his nitrogen solution [Disney] continues to colonize the world – both the imaginary and the real – in the spectral universe of virtual reality, inside which we have all become extras.
>
> (Baudrillard 1996: 2)

> I don't want the public to see the world they live in while they're in the park. I want them to feel they're in another world.
>
> (Disney, cited in King 1981: 121)

Walt Disney's first colonial triumph was the original Disneyland in Anaheim, California. His disdain of the urban development he saw emerging around him is well documented, and much of Disneyland was designed to be an educational and entertaining glimpse at how life could be. Main Street USA, for example, was originally planned to be 'an unambiguous diatribe against the loss of intimacy, the alienation, and the uncontrolled growth of the contemporary city of the car' (Marling 1997: 97). Disneyland would exorcise the ghost of Coney Island, of free-form anarchic pleasure grounds, and, by extension, the unsettling chaos of urban density and the depressing oppression of urban sprawl. It would be frozen in time; it would never grow up to be Los Angeles.

In exploring European possibilities, the Disney Company set out to recolonize the site of their heritage – the theme park was 'going back to its roots, where the early fairs and carnivals from which they sprang, emanated' (Hamnett 1997: 6). This, however, would be an explicitly American version. The first EuroDisney president, Robert Fitzpatrick, explained: 'my mandate [from Eisner] was not to put Mickey Mouse in a French clinic for a face lift. We are going to stay faithful to what we are' (Fitzgerald, cited in Borden 1992: G1). Eisner himself called it 'European folklore with a Kansas twist' (Eisner, cited in Riding 1992: A13).

After considering several European sites, Disney chose Marne-la-Vallée, a primarily agricultural area outside Paris, reportedly because of its centrality to population (over one hundred million live within a six-hour drive), transportation infrastructure, and, somewhat inexplicably, 'good weather' (Riding 1992). Original plans called for the Magic Kingdom theme park proper and a number of surrounding hotels and services, many designed by well-known American postmodern architects such as Michael Graves and Robert A.M. Stern, who incorporated American themes. The Magic Kingdom featured twenty-nine rides and attractions, with most directly modelled after American counterparts.

Thus was set in motion Disney's recolonization project. One of Disney's first tasks, long before the visitor even gets to the gates, is to gently transform the landscape and encourage the visitor to shed all real world associations. Samuel M. Wilson (1994: 28) describes the transformation:

> With the spires of the Euro Disneyland castle looming in the distance like the Emerald City of Oz, we came to the fringes of the park, marked by the first of tens of thousands of newly planted trees. The road wound through a series of landforms that seemed characteristically Disneyesque . . . These disorienting features control the way dramatic views unfold in the created landscape and hide the infrastructure of the huge theme park, with its delivery vans and garbage trucks. We parked half a mile away in the Goofy section of the sprawling parking lots and rode conveyor belts to the entrance, inundated with theme songs. We arrive at the gates, enter the queue, we write our cheques, and prepare to enter a world of fabricated difference, colonial control, economic exploitation, destabilization, and interrogation.

Main Street USA, Adventureland, and other 'discrepant dislocations'

A fundamental postcolonial question focuses on how the West has redefined other cultures both symbolically and materially. In the postcolonial impulse, local geographies are erased and local identities dissolved. The imposition of new economic and political infrastructures and new cultural expectations create 'placeless places' where the possibility of non-Western discourse is replaced with a new essentialized – and non-contestable – version of reality (Emberley 1993, John 1996).

Whatever else they may be, Disney theme parks are the loci of manufactured cultures. Each ride and attraction is an explicit ideological retelling of a cultural narrative. Disney proponents and Disney 'Imagineers' (theme park designers) themselves describe the process as improvement – as programming in all the positive elements while programming out the negative (Fjellman 1992). Disney critics are more apt to reverse the formula and accuse Disney of programming out all the genuine and replacing it with a dangerously simplified conservative agenda. However interpreted, the fact remains that Disney parks create the 'placeless places' of colonial occupation. Disneyland Paris is no exception, and its location in the cultural hearth of the very folklore that Disney redefined adds layers of complexity to the equation. A stroll through the park, with stops at Main Street USA, the Castle, Adventureland, and Discoveryland, can help illuminate the various colonial impulses embedded in the landscape.

At every Disney park, Main Street USA is the first attraction that visitors encounter after entering the gates. The Paris version features images of America consciously repackaged for Europeans, carefully designed to appeal to 'the aesthetic and psychological needs of a European audience' (Marling 1997: 97). These needs were pinpointed, lead Imagineer Eddie Sotto explains, to the 'moment when Europeans first became conscious of what America looked like by going to the movies' (Marling 1997: 97). Unlike the Main Streets at the other Disney parks, Main Street here is allowed to progress beyond the confines of an insular small town. It was intended to have a knowing urban gaze, capturing the transition of a village becoming a city. Clutter is allowed to accumulate around the edges: Sotto mixed popular icons drawn from ads of the period with images from old catalogues and period photographs in order to achieve his blend of Hollywood fantasy and documentary realism (Marling 1997: 99). Ironically, the nuances of Disney's colonial presentation may well be lost on many of the guests. 'Step into an idyllic bygone age as you walk down Main Street USA frozen in time at the beginning of the twentieth century', Disneyland Paris aficionado Phil Hargreaves invites readers in his web site (Phil's Guide to Disneyland Paris 1997). Perusing fan commentary, it is more common to see references to Main Street USA's turn-of-the-century charm than to its cinematic connections. One gets the sense that many guests, already having absorbed Disney's Main Street USA from other settings, literally see what they expect to see.

At the end of Main Street USA sits Le Château de la Belle au Bois Dormant, or Sleeping Beauty Castle. Planning the castle presented an especially delicate colonial situation. The original castle at Disneyland was inspired by Neusch- wanstein Castle in Germany, and the Florida and Tokyo castles are pastiches of the Anaheim one (Borden 1992: G1). To follow their usual mode of operation would have meant repackaging what was stolen, mutated, and turned into a transnational commodity, and even the Imagineers were not up to that level of Euro-pastiche. Instead, designers used a kind of stream of consciousness method, and started with 'the details, the fragments that felt like a castle in this part of France' (Marling 1992: 172). Disney visitors Tom Drydna and Andre Willey report that in their view, the castle is a success:

> Upstairs, the story of Sleeping Beauty is told in ornate tapestries and stunning stained glass . . . Downstairs . . . the Dragon's Lair. The dragon sleeps peacefully next to his pool with the occasional snort of smoke. Then his tail twitches, more smoke, and he begins to awake . . . then you realize his chain is broken! Loved it.
>
> (Disneyland Paris F.A.Q. 1997)

They also loved the inclusion of American commodity forms. 'Don't miss out on the two shops inside the castle, which are charming.'

Adventureland at other Disney parks deserves special mention as the location of Disney's most surreally colonial attractions; Disneyland's Jungle Cruise, for example, features a safari cruise through a tropical-looking setting advertised to be the Nile, Amazon, MeKong, and Irrawaddy Rivers all blended into one. Marling (1997: 111) reports that Disneyland Paris designers consciously tried to upgrade Adventureland for European audiences from something that looks 'plucked from the pages of National Geographic' to 'a knowing meditation on colonial cities where the European-derived conventions of the imperial era have been overlaid by the corrugated iron and crude technology of the early twentieth century' (Marling 1997: 111). How this new sensibility has been translated into themed attractions is not entirely clear; the visitor is still invited into colonial fantasy of Disney's making. 'Pass through Aladdin's exotic eastern bazaar, and follow the beat of the bongo drums into the heart of the Caribbean' (Official Disneyland Paris Web Site 1997). Once there, attractions such as Pirates of the Caribbean, with its 'secret caves . . . skeleton at the wheel in the storm, and the pirate's secret treasure' (as reported by Regan B. Peterson) and Indiana Jones et le Temple du Peril, where Julie Dawe found '1940s-style camp settings, jeeps, tents, etc. that looked like they could have come out of the Indiana Jones movies,' seem to reinforce Disney's tendency to remake local histories and geographies by erasing non-Western identities and replacing them with Western motivations (Disneyland Paris F.A.Q. 1997).

Finally, we reach Discoveryland, a themed area unique to Disneyland Paris. The traditional 'Tomorrowland' slot has been replaced with a 'land' ostensibly dedicated to the visionary ideas of French writer Jules Verne. Discoveryland was clearly Disney's concession to anticipated criticisms of their heavy-handed American presence during the planning process (Riding 1992). Interestingly, rides and attractions in this area are amongst the most popular and the most highly regarded. Though designed by the same stable of Imagineers as the rest of the park, attractions here appear to be more closely aligned with a single literary work and thus less inclined to represent colonial 'improvements' on entire cultural landscapes. Le Visionarium, for example, sends visitors on a 'video trip through time, picking up Jules Verne en route. Very impressive period detail' (as reported by André Willey, Disneyland Paris F.A.Q. 1997). It features an all-French cast, except for H.G. Wells, who is played by Jeremy Irons. Space Mountain (de la Terre à la Lune), similarly, is a wildly popular roller coaster ride based on Verne's book *From the Earth to the Moon*, where as Jean-Marc Toussaint describes, guests are 'pulled into the enormous Columbiad Cannon . . . ricocheted through space, past meteorites, . . . until finally reaching the moon' (Disneyland Paris F.A.Q. 1997).

Yet even in Discoveryland, the result is an uneasy balance between local cultures and Disney's colonial culture. The atmosphere never strays far from Eisner's stated goal of 'European folklore with a Kansas twist.' André Willey's experience with Les Mystères du Nautilus provides a good example: after seeing 'a very authentic-looking Nautilus wait[ing] docked in a pool . . . inviting you to come aboard,' the attraction turned out to be an unimaginative transplant of an American submarine ride. Willey was left wondering 'to be honest, the biggest mystery of all was why is the queue so long?' (Disneyland Paris F.A.Q. 1997). In Discoveryland, we see how the authentic is subtly recentered, recast, and then injected into the colonial narrative. The colonial discourse grows more sophisticated, but also more slippery.

European in blood and colour, but Disney in taste, opinions, morals, and intellect

One of the fundamental mechanisms by which colonial relations with 'othered' people were forged was invitation to mimicry. Local populations were invited, or more properly expected, to rise up to Western value systems, though never to exceed them. That their own culture had integrity was not considered; any cultural redemption was to take place through the curative powers of the West, the sheer breadth of Western management and production of colonial identity, spanning political, social, military, ideological, and scientific thinking (Said 1978).

Disney's production of colonial identity is likewise naturalized, made nearly invisible by its own sheer breadth. Only on rare occasions does it penetrate the surface, such as at the Buffalo Bill Wild West Show where the visitor is invited to 'conquer the West from your dining table' (Wilson 1994: 28). More frequently, employee and guest regulations ensure a more subtle mimicry. American managers required English spoken at all meetings, and American-style security was implemented, including entrance and exit checks (Disneyland Paris F.A.Q. 1997).

The Disney dress code was at the centre of an integrated set of employee regulations designed to Americanize the European workforce on the surface, and, ultimately, would be at the centre of a minor rebellion that forced the Walt Disney Company to revisit the severity of its restrictions. As I discuss in the fourth section, invitations to mimicry can backfire.

Economic exploitations

The third central factor in colonial relationships concerns economic disparities institutionalized through Western controls. Though economic exploitation by a Western country is by definition a necessary condition in 'Third World' status, the argument here is that politico-economic arrangements also have serious cultural implications (Said 1978: 2). A special government office was created expressly to 'make things easier for Disney' (Greenhouse 1991: 6). A master agreement with the then-Prime Minister's government gave the Disney Company the option to develop nearly 2000 hectares surrounding the theme park for leisure, retail, business, and residential purposes in addition to the theme park (Ellis 1991: 38). It was further arranged that Disney could buy the land at prices that were substantially below market value (Riding 1992: A13). The government also agreed to provide improvements to the transportation infrastructure, most notably to extend the Paris Subway to EuroDisney and build a TGV train station 150 yards from the theme park's gates.

The financial arrangements are the third in the set of three interwoven processes that underlie Disney's recolonization project. Together with (re)presenting and speaking for the 'other' and control over employees and guests, this tightly orchestrated economic and political exploitation solidifies Disney's symbolic and material power over the European market. These three factors describe the cultural boundaries of the landscape Disney was trying to create. However, as I will next discuss, the reality of the French theme park often strayed from the intended image. The final postcolonial concern helps to illuminate the complex interchanges between French, European, and American; between visitor, employee, and general public; and between corporation, nation, and locality that undermined the stability of Disney's project.

Hegemonic destabilization

Discrepant dislocations . . . produce unintended effects.

(John 1996: 16)

One of the most fascinating aspects of postcolonial analysis is the recovery of agency, and with it, the unpredictability of non-linear histories. As Barrett (1996:

243) argues, 'postcolonialism refutes the determination of colonization purely in terms of monolithic brute force, by pointing instead to colonialism's conflicting demands for stability and reform, and by exposing the contradictory agendas of the various agents of colonialism'. He opens up crucially important theoretical ground, because the adoption of such a stance allows one to question formulations of power in new ways. Drawing most commonly from critical theorists including Bahktin, Foucault, and Gramsci, postcolonial writers look at how the interaction between 'us' and 'them' 'simultaneously stabilizes and destabilizes colonial authority' (Lee 1997: 92). While Lee (1997: 95) later cautions against re-interpreting history and projecting resistance back onto the past by 'reconstructing a space for critical activity' when in fact none was there, the general theoretical impulse remains vital for sketching in the boundaries of colonial relationships, and, in particular, understanding the fate of Disney's attempted recolonization from the beet fields.

In Disneyland Paris the well-tooled machinery of Disney Obedience lurched to a stop. The weight of cultural conflict became too great for the machine to bear: employees, visitors, and general public all participated in this unrehearsed drama. Even before the EuroDisney park opened, the Disney Company discovered themselves embroiled in rounds of cultural negotiations that they had not anticipated. Perhaps the problems long predated Disney's entrance: Robinett and Braun (1990: 18) comment that the French were 'bemused by, and unaccepting of, some of the fundamental principles of theme park operation.' The prohibition of home-prepared picnics, for example, sent a message as negative as the high-priced admission. Olivier de Bosredon, head of Asterix Theme Park, also in France, was led to the conclusion, as he watched his park teeter on the edge of bankruptcy, that 'the average French person has no need for an amusement park . . . In France, leisure remains very Traditional' (de Bosredon, cited in Theobald 1991: 2).

Relations behind the scenes were strained. European employees balked at the restrictions. Eventually it became clear that the park would have to ease these, which was in itself rather out of character for the company; Disney's previous refusals to negotiate on American labor matters had provoked bitter strikes (Smith and Eisenberg 1987). A central focus became the more explicit incorporation of European cultural sensibilities. A French president, Bourguinon, replaced Fitzpatrick. One of Disney's first hegemonic concessions was to lift its ban on alcohol, which had been of great concern to European employees and guests alike. Disneyland Paris is now the only Disney theme park that allows alcohol within its gates, though to date it is only served with meals at full service restaurants (Secrets of Disneyland Paris 1997). Employee training sessions have likewise been relaxed. Larcomb reports receiving advice that would have been utterly shocking at an American 'Disneyland University'; her instructor advised the class

'If you don't feel like smiling, don't' (Secrets of Disneyland Paris 1997). Apparently French employees took this training advice to heart and even improvised on it: Andrew Sutton and Susan Williams, in an otherwise complimentary web page devoted to Disneyland Paris, commented of their vacation to the theme park, 'some French cast members were friendly (a little TOO friendly!) and others are more grumpy' and concluded 'the park compared very well to the other versions, but the staff did not. The French, bless 'em' (Sutton and Williams 1996). Likewise, Margie, a visitor from Missouri USA, found 'it's a very different service attitude. It's a big problem. You're serving them' (cited in Allen 1994: 17).

Such changes to Disney policy were the culminations of complex hegemonic interplay between Disney management and employees, colonizer and colonized, that was carried out over a period of time by various actors. It is important to note that no single path of action proved successful in itself: the process of cultural change at Disneyland Paris ultimately included a combination of the organized and the sneaky, the politically active and the rebellious. We can return to the subject of Disney's dress code outlined above in order to observe this process in closer detail.

The Disneyland Paris dress code directly conflicted with French cultural standards regarding dress and grooming, and this quickly became a source of friction between the Disney Company and French citizens, both employees and the general public. Some potential employees refused to complete the application process when they learned of the restrictions, such as the musician who chose to keep his pony tail rather than become a member of the Disney jazz band (Disney Dress Code 1991: 48). Others who were hired later mounted organized campaigns against certain aspects of the dress code. Female workers, for instance, eventually won the right to wear red lipstick and black panty-hose by arguing, among other things, that even Minnie Mouse was allowed charcoal-coloured tights (Theobald 1991: 2). Disney's restrictions were generally ridiculed in the press, and French labor unions protested against what one group called Disney's violation of 'human dignity' (Disney Dress Code 1991: 48).

The conflict entered into the legal realm when a formal complaint was lodged against Disney by a government labor inspector. The details of the complaint reveal that the conflict was not simply about style of dress: it centred more fundamentally on the nature of the power wielded by an American multinational corporation. Disney presented the dress code as an attachment to the employee contract; as such, it was exempt from standard approval through government channels. The French government argued that in fact the regulations constituted a disciplinary code applicable to all employees, and therefore under French labor law should have been written directly into the contract and hence subject to proper approval. Disney eventually backed down, and made adherence to the regulations voluntary. As Disneyland Paris employee Liz Lancomb reports, the ensuing atmosphere was looser than at the American parks, where she has also worked.

> I was not required to wear my hat, scarf, or belt, tuck in my shirt, even wear black shoes. We were asked to, but many did not and could not be punished for it. Once in Florida [Walt Disney World] I was sent back to costuming for the nail polish I had on, while in France anything goes. A friend of mine who worked at Camp Davy Crockett [in Disneyland Paris] even had a nose ring!
>
> (Secrets of Disneyland Paris 1997)

Another important voice in this hegemonic discourse has been that of park guests. The perceived imposition of American values, as with employees, has provoked concerted protest that eventually led to modifications by Disney management. Language has been an especially volatile issue, and in particular the policy Disneyland Paris has towards multilingualism. The theme park officially recognizes six languages: English, French, German, Italian, Spanish, and Dutch, though in practice only four (English, French, German and Italian) are commonly encountered. Initial attempts at linguistic diversity resulted in an uneasy blend of imperial English-only and surreal polycacaphony. A truly multilingual environment seemed to elude the park's technological capacities. A patron who visited the Le Visionarium attraction discovered 'the headphones are useless! I had to keep fiddling with it since no matter if I wanted to hear English translations, no matter which button I pressed, I got German in one ear and Italian/English in the next' (Secrets of Disneyland Paris 1997). Visitor James Bohn found that live humans were not much of an improvement, noting that most employees know only a handful of key statements in their non-native languages and have not learned the polite forms. Thus, English-speaking guests might be greeted with a brusk 'Off!' rather than a 'Please step out to your right' upon exiting a ride (Disneyland Paris F.A.Q. 1997).

More problematic, though, was Disney's choice of language in prominent attractions. Predominantly English-oriented attractions were the most heavily criticized. The soundtrack for Phantom Manor, for example, was originally recorded by Vincent Price in English and French. But due to complaints from French guests, within five months the tape recording was replaced by a French-only narration recorded by a French actor (Secrets of Disneyland Paris 1997). For some guests, any language that was not French was considered a cultural affront. Attractions that featured recordings in other languages, such as Snow White (German), Pinocchio (Italian), and Peter Pan (English), also received complaints. They, like Phantom Manor, were all converted to French-only narrations (Secrets of Disneyland Paris 1997).

121

Linguistic representation, however, is inseparable from cultural representation; Disney's translations sometimes led to a degree of cultural schizophrenia. Reporter Julia Llewellyn Smith braved a rainstorm to visit the Tom Sawyer attraction, which she concluded was oddly incongruous: 'the plummy Gaellic voice droned on with implausible accounts of his life as a steamboat captain and his vieil ami Tom Sawyer. The crowd, consisting largely of English-speaking Germans, Italians, Belgians, and Slovenians yawned' (Llewellyn Smith 1993: 13). Children, especially European ones not yet steeped in global Disney culture, likewise can find Disney's versions problematic. Samuel M. Wilson, anthropologist and father (1994: 30), visited Sleeping Beauty's castle and found that 'as we milled about in the crowd, we saw parents from all over Europe making quick interpretations apparently trying to connect the Disney characters and events to those their children knew.' He later fell into the same predicament in reverse when touring the real Sleeping Beauty's castle, Château d'Usse (where Charles Perrault was inspired to write the story), where he 'had a lot of explaining to do' to convince his four-year-old daughter that this Sleeping Beauty was not an inferior Disney imposter.

A final form of visitor interpretation involves the wholesale intentional hijacking of meaning. This can be the most subtle, yet also most disturbing; it most often seems born of violent impulses. Pirates of the Caribbean, for example, provided a sinister forum for a group of teenage boys observed by journalist Jennifer Allen. All visitors who exit the ride are funneled through a gift shop featuring pirate-like merchandise – 'swashbuckling sword, 30FF; captain's hook, 20FF' (Allen 1994: 16). The three boys were gleefully pulling shotguns off the racks. '"Christmas presents?" I ask. "Chouette," one says, meaning "hip," as he holds the barrel up against the temple of his friend's head. The boys are here with their families on a military discount' (Allen 1994: 16).

For some guests, as well as members of the general public, merely the presence of a Disney logo inspired derisive comments and sometimes violent outbursts. When the park first opened, 'characters' (employees in costume as Mickey, Minnie, Goofy, and other Disney cartoon likenesses) risked being beaten up by park patrons. For several months, guests and employees alike were scanned with metal detectors at the park entrance after an employee was cut in the face by a patron carrying a razor blade. Employees who wore their 'EuroDisney Opening Crew' jackets outside the park were also popular targets in downtown Paris (Secrets of Disneyland Paris 1997). In an effort to reduce attacks, Disney has announced it will provide employees with miniature cameras that will be concealed inside costumes. These little cameras, significantly, will have the ultimate effect of monitoring not just unruly guests, but also employees themselves (Letts 1996: 11).

Conclusion

In the end, according to recent financial reports, Disney's recolonization project has at last proven successful. It attracted 11.7 million visitors in 1996. Disneyland Paris is making a profit, and it is now the most popular tourist attraction in France. Disney management appears vindicated in their conviction that the project was a worthwhile endeavor. Indeed, the theme park represents an extremely worthwhile endeavor, bearing as it does important commentary about Disney's relation with the world, but perhaps not at all in the way the Disney Company had intended.

Disneyland Paris has been a site of postcolonial struggle since its inception, albeit a site where struggle occurs within and amongst different cultures of power. Some might consider it divine justice that the twentieth century, which began with much of Europe still in a position of colonial domination, should end with Europe being recolonized by one of its former colonies. Yet the ensuing 200 years since America gained its independence have brought about fundamental economic, social, and political changes. Europe finds itself victim of a new form of colonial power: the transnational corporation. In fact, Disneyland Paris speaks to a fear lurking deep in social theory: that the nation-state is obsolete, about to be eclipsed by the multinational corporation.

It is at tourist destinations like Disneyland Paris where, beneath their cloaks of fantasy and frivolity, these new relations are being worked out and contested. We have seen that it would be theoretically unwise to homogenize postcolonial experience into an active 'us' and a passive 'them'; instead, the most fruitful observations can be made at the hegemonic balance between the two poles. On the one hand, we can uncover indisputable evidence from the theme park world that multinational corporations are gaining a stranglehold on European life. A number of Disney attractions, for example, are sponsored by corporations such as IBM and Kodak, and other American corporations including Anheuser-Busch and Warner Brothers are purchasing other European theme parks. But on the other hand, we find equally indisputable evidence that local populations, through a variety of postcolonial cultural practices, can successfully challenge the supremacy of corporate culture. The fact that there is now a Disney park populated by grumpy women in red lipstick and black tights suggests that as we enter the twenty-first century the structural conditions underlying postcolonial relationships may be undergoing a sea change, but the fabric of postcolonial debate remains richly intriguing.

References

Allen, J. (1994) "Euro Disney Postcard: The Tragic Kingdom," *The New Republic* 4, 121, 122: 16–17.

Barrett, J. (1996) "World Music, Nation, and Postcolonialism," *Cultural Studies* 10 (2): 237–47.

Baudrillard, J. (1983) *Simulations*, New York: Semiotext(e).

—— (1996) "Disneyworld Company," *Ctheory Electronic Journal*, no. 25, 27 March 1996.

Blake, P. (1972) "Walt Disney World," *Architectural Forum* 136: 24–31.

Borden, L. (1992) "Disney, Euro-Style," *Wilmington News Journal,* 29 March: G1.

Cohen, R. (1992) "Resisting Disney: Unmitigated Gaul," *New York Times* 9 April: D1.

"Disney Dress Code Chafes in the Land of Haute Couture" (1991) *New York Times*, 25 December: 1.

"Disneyland Paris F.A.Q." (1997) Web Page: http://www.informatik.tu-muenchen.de/~schaffnr/etc/disney

Doss, E. (1997) "Making Imagination Safe in the Fifties: Disneyland's Fantasy Art and Architecture," in K.A. Marling (ed.) *Designing Disney's Theme Parks*, pp. 179–90.

Eco, U. (1986) *Travels in Hyperreality*, Harcourt Brace Jovanovich: San Diego, New York, London.

Elliot, H. (1993) "Disney Park Bows to European Tastes," *The Times*, 30 March: 3.

Ellis, C. (1991) "Disney Goes to Paris," *Landscape Architecture* June: 38–40.

Emberley, J.V. (1993) *Thresholds of Difference: Feminist Critique, Native Women's Writings, Postcolonial Theory*, Toronto, Buffalo, London: University of Toronto Press.

"Euro Disney Growth to Lag" (1994) *New York Times*, 1 April: D14.

Fjellman, S. (1992) *Vinyl Leaves: Walt Disney World and America*, Westview Press: Boulder, San Francisco, Oxford.

Fogelsong, R. (1990) *Magic Town: Orlando and Disney World*, Graduate School of Architecture and Urban Planning, UCLA: Los Angeles.

Frankenberg, R. and Mani, L. (1993) "Crosscurrents, Crosstalk: Race, 'Post-Coloniality', and the Politics of Location," *Cultural Studies* 7(2): 292–310.

Greenhouse, S. (1991) "Playing Disney in the Parisian Fields," *New York Times,* 17 February: Section 3: 1.

Hamnett, J. (1997) *EuroDisney: A Cross-Cultural Communications Failure*, unpublished manuscript, Anderson School of Management, UCLA. Available on the Internet http://www.anderson.ucla.edu/research/conferences/scos/papers/hamnett.htm

Harrington, M. (1979) "To the Disney Station," *Harper's*, January: 35–44.

John, M. E. (1996) *Discrepant Dislocations: Feminism, Theory, and Postcolonial Histories,* University of California Press: Berkeley, Los Angeles, London.

King, M. (1981) "Disneyland and Walt Disney World: Traditional Values in Futurist Form," *Journal of Popular Culture* 15(1): 116–40.

Lavie, S. and Swedenberg, T. (1996) "Between and Among the Boundaries of Culture: Bridging Text and Lived Experience in the Third Timespace," *Cultural Studies* 10(1): 154–79.

Lee, K.-W. (1997) "Is the Glass Half-Empty or Half-Full? Rethinking the Problems of Postcolonial Revisionism," *Cultural Critique* 36: 89–118.

Letts, Q. (1996) "Snow White Gets Mini-Camera to Keep Bodice Pure," *The Times*, October: 11.

Llewellyn Smith, J. (1993) "What Price a Wet Day Out at Disney?," *The Times*, 18 August: 13.

MacKay, D. (1997) Personal Communication, 4 October.

Marcus, G. (1997) "Forty Years of Overstatement: Criticism and the Disney Theme Parks," in K. Marling (ed.) *Designing Disney's Theme Parks: The Architecture of Reassurance*, pp. 201–8.

Marin, L. (1984) *Utopics: Spatial Play*, Humanities Press: Atlantic Hughlands, NJ.

Marling, K.A. (ed.) (1997) *Designing Disney's Theme Parks: The Architecture of Reassurance*, Canadian Centre for Architecture: Montreal.

—— (1997) "Imagineering the Disney Theme Parks," in K.A. Marling (ed.) *Designing Disney's Theme Parks: The Architecture of Reassurance*, 29–178.

"Official Disneyland Paris Web Site" (1997) Web Page: http://www.disneylandparis.com/

"Phil's Guide to Disneyland Paris" (1997) Web Page: http://www.seekfuture.co.uk/dlp/

Phillips, A. (1993) "Where's the Magic?," *Maclean's* 106(18): 47.

The Project on Disney (1995) *Inside the Mouse: Work and Play at DisneyWorld*, Duke University Press: Durham, NC.

Riding, A. (1992) "Only the French Elite Scorn Mickey's Debute," *New York Times*, 13 April: A1.

Robinett, J. and Braun, R. (1990) "French Theme Parks: A New Industry Travels a Bumpy Road," *Urban Land* 49(9): 15–19.

Rodriquez, N. (1997). Web Page: http://www.cs.uoregon.edu/~nrodriqu/disney.html

Ross, A. (1988) *Universal Abandon: The Politics of Postmodernism*, The University of Minnesota Press: Minneapolis.

Said, E. (1978) *Orientalism*, Vintage: New York.

"Secrets of Disneyland Paris" (1997) Web Page: http://www.oitc.com/Disney/Paris/English/Secrets.html

Smith, R.C. and Eisenberg, E.M. (1987) "Conflict at Disneyland: A Root-Metaphor Analysis," *Communication Monographs* 54: 367–80.

Soja, E. (1989) *Postmodern Geographies*, Verso: London and New York.

Sorkin, M. (1992) *Variations on a Theme Park*, The Noonday Press: New York.

Stephanson, A. (1988) "Regarding Postmodernism – A Conversation with Fredric Jameson," in A. Ross (ed.) *Universal Abandon*, pp. 3–30.

Sutton, A. and Williams, S. (1996) Web Page: http://www.karoo.net/altair/france/edisney

Taylor, J. (1987) *Storming the Magic Kingdom: Wall Street, the Raiders, and the Battle for Disney*, Ballantine Books: New York.

Theobald, S. (1991) "Why Do the French Find the Magic Kingdom so Galling?," *World's Fair* 11(4):1–2.

Warren, S. (1994) "The Disneyfication of the Metropolis and Popular Resistance in Seattle," *Journal of Urban Affairs* 16(2): 89–107.

—— (1996) "Popular Practices in the Postmodern City," *Urban Geography* 17(6): 545–67.

Wilson, S. (1994) "Disney Dissonance," *Natural History* 103(12): 26–31.

Zukin, S. (1995) *The Culture of Cities*, Blackwell: Cambridge, MA and Oxford, England.

9

NOMADIC-SYMBOLIC AND SETTLER-CONSUMER LEISURE PRACTICES IN POLAND

Ada Kwiatkowska

Introduction

This paper considers the negotiation of leisure practices and knowledges in the current flux of Polish culture and everyday life. Poland's contemporary leisure has distinctive cultural influences and forms. These relate to cultural tradition, to a political concept of mass-consumption of the recent past and to the concept of individual consumption of Western civilisation. There is considerable flux in these patterns due to socio-cultural transformations in Central European countries. Leisure practices are considered according to their symbolic meaning and spatiality in two categories: nomadic-symbolic, pilgrimages, secular and religious holidays, and settler-consumer: private gardens, weekend-houses, feasts.

Leisure contexts

Leisure can be less a feature of real time than a concept relating to the quality of time based on a social hierarchy of values, values that determine a direction of cultural development, situating human actions within 'a temporal framework of laws and conventions' (Parkes and Thrift 1980: 73). Particular ideas, beliefs (mythology, religion, theology) and human knowledge (science, philosophy, art) have influenced Polish leisure in distinctive ways, and do so in new hybrid forms. These influences have been religious or political doctrines and mass media propaganda. Religious ideologies create many visions relating to the quality of space-time, which are reflected in notions of leisure. Vision of paradise, a beautiful place of perfect happiness, where people are free from work pressure (Delumeau 1996: 7–10) is contrasted with living on the earth, where people are sentenced to work hard. The biblical Eden garden is an origin of the paradisaic leisure or ersatz paradise, often used in tourist promotion.

A vision of soul-journey from birth to death, from resurrection to immortality, expresses the importance of journey as a time of growing older, self-perfection, experiencing temptation of the material world or telling the truth. Like *karma* this means the sequences of action and reaction, and different categories of consciousness (Kapleau 1988: 315, 325). Such a soul-journey has a spatial representation in the form of ritualistic path, dance paths during initiation ceremonies (Rykwert 1982: 106), the path of the Cross. The meaning of the journey is a time of trial and experience of the beginning and the end, as a path connected with rituals and signs, and as a challenge to self-perfection. In a concept of daily routine (the world created in six days, the seventh a day of rest) two qualities of time are defined. Work is a secular time connected with creating, producing or gathering the material things to satisfy the needs of body; leisure, sacred time connected with praying, meditating or spiritual self-perfection and concerned with the needs of soul. Modern leisure patterns are usually in conflict with religious celebration, yet hold expectations of wonders and searches for the intensive, exciting, spiritual experience in leisure time.

Political ideologies express the socio-economic value of work and leisure time. They define workers' rights, eight hours a day, four weeks vacation. A proscribed work process is based on economic values relating to market rules, e.g. profits, productivity, efficiency, supply and demand, and to social policy, division of labour, profits, private and public property (Marx 1951: 77–80, 177). As leisure is treated as supplementary to work the same terms are used. The political influence on the idea of leisure is considerable. The questions dividing the political scene into right and left sides, relate to work and leisure (Habermas 1983: 449–74). The social control of leisure patterns, leisure as private or public sector, part of privileges or human rights, as tax-free or tax-luxuries, leisure equality, the limits of freedom, are distinguished by political ideologies.

Mass media propaganda influences leisure practices directly, because the media constructs powerful representations reflected in people's imagination, 'the Nemesis of creativity' (McLuhan 1964: 68), dictating needs and dreams, market rules generating desires for new goods, constructing distinction, selling leisure in 'brilliant packages' (Kwiatkowska 1996: 1–2), reconstructing people as consumers who 'buy . . . the illusions, signs and symbols' in the world of 'simulated hyper-reality' (Dickens 1994: 379–81).

Different forms of social organisation deploy different rules, norms and customs which govern and regulate leisure patterns and utilise a variety of customs and forms. Leisure can be institutionalised or an activity of informal groups. This difference results in different uses of space, exemplified in a transpatial community characterised by mechanical solidarity with ritualised relation between members, using bounded inner space (guild, communitas) or spatial community characterised by organic solidarity with virtual relations, using unbounded outer space (neighbourhood, crowd) (Hillier 1989: 18). In the spatiality of leisure practice, transpatial communities are represented in institutionalising forms of leisure interests as clubs (golf, caravanning, yachting, hiking, wildlife exploring etc.) with their own network of relatively detached and bounded meeting places. Spatial communities are informal groups or individuals (tourists) meeting together accidentally in a certain place which focuses their interests.

Leisure practices are shaped by identification with different social groups, e.g. family, friends, professional groups, hobby groups, age-groups, society, nation. Leisure practices as life*style* comprise social references, orientation and temporality (Sicinski 1977: 299). They also comprise socio-spatial mobility. Migratory social patterns may be connected more with space than with place, space as a physical, mental or poetic challenge which may emerge in contemporary Western caravanning (Wallis 1993:425). Modern leisure emerged as time away from production, differently structured from holidays linked with traditional cultivation or growing, changing the perception of space-time (Kwiatkowska 1997: 407) in leisure, influenced by cyberception and hyper reality (Titman 1995: 49, Rojek 1995).

Individual practices relate to values and desires. Materialistic/consumptive values may be distinguished where 'to have' is more valuable than 'to be'. In spiritual and non-consumptive values 'to be' is more valuable than 'to have' (Fromm 1976). These are reflected in leisure choices. Another cultural distinction concerns different meanings of activity and non-activity. Action signifies 'doing' in active (Western) culture and 'being' in non-active (Eastern and Southern) culture. This distinction expresses active (dynamic) and passive (static) relations with space-time. In active culture, action realises itself in sequences of space-time. Waiting or sitting, seemingly a lack in space-time, mean lack of activity (doing nothing, loss of time). In non-active culture, action is not connected with moving along space-time, therefore waiting or sitting can signify the activity (doing something, e.g. thinking, perceiving, time winning) (Hall 1987:153). This difference is reflected in individual leisure patterns, respectively orientated to 'doing' as 'to go', 'to travel', 'to explore', or to 'being' as 'to feel', 'to understand', 'to taste'.

Polish culture and its influence on leisure practice and knowledge

Central Europe sits between Western and Eastern cultures, on the margin of both. Each expresses different ideological systems and values. Whilst constantly changing, each persisted in opposition: Christian and pagan, Roman-Christian and orthodox worlds, secularism and fundamentalism, democracy and absolutism, individualism and state supremacy. Being on the porous borders of cultures makes each unstable. They are transformed into new qualities of provincial living. New ideas come later or in hybrid form. This section considers the consequences of this cultural distinction in terms of cultural constituents already outlined.

Polish mythology is shaped by the romantic vision of missionary resulting from its specific geographical position and history connected with fighting for independence (Zaluski 1977:160–4). Its position changed from an Eastern to Central European country over the centuries with the change of its borders, but the Polish mission as 'bulwark of Christianity' has survived. Poland protected Europe against the Eastern orthodox world or, later, communistic expansion, often protecting as the loser.

The Roman-Christian Church was the fundamental basis of Polish culture. It played a very important role when Poland lost independence, its power both symbolic and real. Its culture continued based on national and religious mythology and rituals, which sustained the dream of independence (Stepien 1989: 208–82). The religious holidays, such as Christmas or Easter, became nationally cel-ebrated at different times than in the Orthodox Church. This alliance of Polish and religious culture was strengthened after the Second World War when com-munist propaganda tried to eliminate Polish and Christian traditions and to create new socio-realistic, international and atheistic culture. However, the alliance became stronger. Now the conservative part of the Polish Catholic Church sustains a myth of Poland as a bulwark of Christianity. Paradoxically, contemporary cultural instability seems to come from the Western laicised civilisation.

Religious symbolism and rituals survived. They contextualise leisure in the form of Christian holiday celebration, participation in weekly mass, in pilgrim-ages, and Sunday observance is strictly practised. In opposition to Western Christianity, which is characterised by the discursive relation to religious doctrine, Polish Christianity is very dogmatic, formal and ceremonial. Participation in the practices of mystery is highly regarded and incorporates distinctive pagan beliefs and rituals connected with the outdoors that influence ideas of 'nature'. Slavonic tribes adored the sacred trees, animals or giants and holy places, e.g. a moun-tain, glade or lake, sites of annual pilgrimage (Gieysztor 1982). Slavonic rites co-exist in the 'Spring welcome' and 'St John's Night'.

Contemporary practices are still influenced by the political ideology of the recent period. The socialistic system created and propagated the concept of 'working man' [sic]. The state organised the work time but also took care over leisure time. The state propagated the idea of collective recreation. Most of the state-owned factories and institutions had holiday centres for their workers. They organised excursions, wanderings, going mushrooming, feasts etc. for workers and their children. They also distributed the workers' garden lots. Holiday centres, such as boarding-houses, rest-houses, cottages or sports facilities subsidised by the state are now transferred to private ownership.

Mass media, powerful in promoting collective interests and participation of the working class in high culture, 'educated', without advertisements (Kloskowska 1980: 470–2). People were subjects of consumption and not an object of market play, thus conscious of their own needs, ironically conveyed through the mass media. This included visual representation of 'happy children' being in camps organised by the public schools or 'happy coal-miners' sunbathing on the beach in the holiday centres of the state-owned mines during summer vacations. It signified the care by the state of the working class; the attraction of collective recreation, that everyone had easy access to holiday centres. The last decade of Polish transformation has brought a new language of mass media propaganda. Initially people were defenceless in confronting the commercial world, and are now more conscious of themselves as active consumers.

Migration from villages to cities either side of the Second World War resulted from industrialisation and changing borders. In 1946, 70 per cent of the population lived in the country, now this is almost reversed (Statistical Yearbook 1992: 26). Many people living in the cities now are first generation, their connections with the country very close. Country customs and country tradition remain significant in Polish culture. In consequence there is great vitality in religious holiday practice and strong family participation.

Ideologically leisure was significant during the communist period, especially for young people. Scouting and the Union of Polish Youth were very active. The contents and context of play, amusement, wandering or camping became important practices to influence young people separated from their families. Polish propagandists were never so successful as their colleagues in East Germany, where the quasi-military organisation of FDJ (die Freie Deutsche Jugend) seemed to control all youth leisure activities. Yet the institutionalisation of leisure activities required that every factory, office or holiday centre employed someone responsible for the cultural and educational programmes of leisure time (Jankowski 1977: 42–4); to organise feasts, dancing, wandering, to buy theatre or opera tickets (expressed in Voyage, a satirical film by Marek Piwowski).

There is a long tradition of social activity in informal groups, cultivated during times of cultural struggle and self-help and self-generation. Leisure activities amongst groups of friends expressed something more than a wish of meeting, playing or being together. They extended individual freedom and space in which people escaped institutional control, accompanied by the atmosphere of illegality and mystery. Now especially the young seem to look for mystery in informal quasi-religious groups (e.g. Emanuel, Hare Krishna) or illegality in subcultural groups (skinheads, punks), rejecting institutionalised social activity and materialistic, commercialised mass culture.

Unlike the experience of other Western countries Poland experienced a late transformation to modern patterns of leisure and work connected with a socialistic construction of producers not consumers. A five-day week to secure good production was realised after many other countries. However, the creation of the service infrastructure – banks, hotels, gastronomy, tourist agencies – remains to be done.

Polish leisure activities are contextualised by both the patterns of active culture of the West and non-active culture of the East. The state socialistic system strengthened non-consumptive values as a reaction to the lack of desirable consumer goods. Habits of passive participation and expectations of a protective state and an uncritical adoption of Western leisure practices directed towards individual consumption have been shown to disturb contemporary cultural transformation. This emerges as caricature, with an uncertainty of when to have, to use; to see and to be seen perhaps as more important than to be, to create, to understand or to have rest. Leisure time activities can be regarded as a kind of competition of display, who can pay more or travel farther.

Leisure and spatial knowledge

Leisure time activities express themselves in space-time and in the spatial settings where they are realised. Two categories of leisure, nomadic and settler, provide a means to organise and to consider these processes. 'Nomadic' practices involve a sequence of places, characterised by mobility and passing space-time, by temporary and changeable settings. Settler leisure relates to one place, characterised by the stationary, permanent and fixed settings (Barbey 1993: 109).

These differences are significant and may be used to distinguish emerging leisure practices culturally. Thus, we can signify nomadic practices as expression of the myth of the voyage, i.e. the distance between the home and the world where 'ego alone resides' (Dal Co 1990: 21). The settler pattern acknowledges the myth of the centre of the world, a reference system and the 'illusion of permanence' (Barbey 1993:109). The latter is connected with an exchange of ideas and goods, the latter with production and consumption. The nomadic relates to integrated social groups with ritualised relations between members, identifying themselves with the group through common values and a transpatial practice. Settler activities require an integrated socio-spatial group with an organic or virtual relationship between the members; whose identity is formed on the basis of shared neighbourhood or practice. Breaking the barrier of space-time by mobility signifies nomadic practice and by virtual communication increasingly characterises settler practice. The two are again contrasted as the 'nomadic' intensifies the experience of 'being' – because of the limits of mobility in space-time and is essentially symbolic, whilst the 'settler's' experience is of 'having' – because of the potential of place constituted of consumption.

The distinction between these categories relates to different ways of realising leisure practices. The following section interprets these distinctions using interview and quantitative evidence detailed below.

Nomadic

Nomadic activities, expressed in passing through space, are generated by the symbolic meaning or aesthetic value of place (sanctuaries, landscape monuments) and symbolic meaning of time that prompts mobility (Saints' Days, Christian holidays, First Day of Spring, May-party, gathering the forest fruits). Present-day activities relating to symbolic places are dominated by religious activities. Pilgrimages to sacred space, especially to Czestochowa, religious capital of Poland, were always part of the Polish landscape, but previously only amongst village dwellers. Due to the recent Roman-Christian ideological expansion there is a growing phenomenon of religious tourism, in which the members of all social groups take part. There are 382 sanctuaries in Poland (Krezel 1983: 78). Pilgrimages are socially distinguished, professional groups: businessmen, teachers, workers, farmers; gender and family groups; age-groups, students, pensioners. The liturgical ceremonies are similar but spiritual strength is seemingly focused through a company with a common basis of identification. The most popular pilgrimage, to Czestochowa, lasts 5–15 days and starts in every diocese. Pilgrims walk about 30 kms a day, praying and meditating (Radecki 1990: 50–8). Pilgrimage is a time of penance, limitation of needs to the minimum, sharing bread with other people. Clothing and food, sleeping-bags and tents are transported by cars. Pilgrims take with them only necessary things. Stops are organised along the roads, to have a rest, to eat and to pray. Pilgrims usually camp out in the fields, sometimes using local hospitality. The scale of this phenomenon has prompted local commercialisation.

The socio-cultural transformation of Poland has not as yet changed the traditional perception of landscape as eternal value and elation, a mythological setting of childhood. People practice these attachments in increasingly diverse ways. Particular archetypes of Polish landscape are linked with aesthetic and tourist attractions: Tatry's mountains, fashionable for winter holidays and climbing; Pieniny, picturesque landscape; Hel, a fashionable summer holiday centre, swimming and life enjoyment.

The most popular form of recreation is two weeks holiday organised by employers or tourist agencies in holiday centres (in 1991 involving 22.5 million people). Tourist interests in active forms of recreation, such as cycle tours with a group of friends, until recently very popular, are decreasing. Participation in touring declined over ten years from 1.6 to 0.6 million participants (1991), cycling declined from 100,000 to 40,000, sailing halved to 30,000 over the same period (Statistical Yearbook 1992: 452). Higher environmental standards are expected. However, the data omits casual participation, and there may be a significant shift in the way people do this.

Many dates in the Polish calendar are connected with symbolic meaning and still prompt mobility. Many of these express a sacred dimension, 'the other'; some are secular. All Saints' Day commemorates the dead, a time of meeting with one's own history. The past, written on the graves of cemeteries, plays a very important role in Polish culture. Post-war migrations and border changes have meant that many people have their families' graves far from home and so have to travel to take care of the graves and for family reunions. Other Christian holidays are connected with the meeting of the whole family, perhaps three generations, and are times of great mobility. In residual pagan celebrations many young people travel and wander through the countryside (St. John's Night (June 21/22)). This is a time of life affirmation (Zygulski 1981: 211–13, Gieysztor 1982: 206–7). 'May-party' celebrates a workers' tradition of helping farmworkers during the weekends, organising a party in the countryside, cultivated in the socialistic period, but now of little attraction.

Gathering the forest fruits is one of the most popular leisure activities in Poland. Gathering mushrooms and berries is practised in the family or with friends. These also nomadic activities construct their own spatiality, using transport facilities (trains, buses, cars), networks of paths and tourist attractions, gastronomic facilities and night rests (hotels, hostels, motels, camping).

Settler-consumer

Consumption and sedentary (home-based) leisure have accelerated in popularity to become almost as popular as more traditional nomadic practices (Kwiatkowska 1994: App.1). Incidental, seasonal activities, 'folk culture', such as circus, feasts and funfairs retain popularity. Refiguring leisure emerges again in terms of settler practices, and can be exemplified in two examples, gardening, less directly commercialised than other examples, and second-homing. Gardening has mixed popularity, and one third of households have a garden, and more people have a garden lot (allotments). These lots provide an insight into the knowledge of leisure practice. In the socialistic period garden lots were financially crucial and were allotted as workers' privileges. Allotments remain a popular settler-consumption leisure of lower and middle class. Gardens are increasingly valued for their aesthetics and reflective constituents, ironically continuing the value of allotments from the previous period, when they enabled individual expressiveness as well as easing food shortage. Weekend-cottaging, previously subject to disruptions through border changes and migrations, became fashionable in the 1970s. Village populations grew old, their children lived in cities. They sold their houses because there was no family to take care of the house and land, and many were in poor repair. Those who bought these houses spent a lot of money renovating and adapting them to their needs. These reasons contribute to both the continued attachment to settler site-based rituals in a rapidly modernising culture and to their adaptation.

Conclusions

The significance of cultural influences is evident in many contemporary leisure practices and in their shifts over recent history, contrasted in the socialistic concept of the worker and collective recreation; the religious concept of meditation and collective pilgrimages; the Western concept of 'self-perfection' and collective alienation. Negotiation in order to realise senses and meanings in leisure practices in relation to these contexts has always been significant, now in terms of competing proffered choices and desire to resist. Contemporary leisure practices and the meanings that particular sites of practice hold are in acute flux and are being negotiated. The expansion of Western mass media propaganda in advertisements, consumer goods and tourism, the unification of the European market and their easy access provoke a Westernisation/commodification of leisure practice and the reinvention of leisure sites, tempting myths of self-realisation and expression.

References

Barbey, G. (1993) 'Spatial archetypes and the experience of time: identifying the dimensions of home', in E.G. Arias (ed.) *The Meaning and Use of Housing: International Perspectives, Approaches and Their Applications*, Avebury: Brookfield, Sydney, Aldershot.

Dal Co, F. (1990) *Figures of Architecture and Thought: German Architecture Culture 1880–1920*, Rizzoli International Publications: New York.

Delumeau, J. (1996) *Historia Raju* (orig. *Histoire du paradis*), trans. E. Bakowska, PIW: Warszawa.

Dickens, P. (1994) 'Alienation, emancipation and the environment', in S.J. Neary, M.S. Symes and F.E. Brown (eds) *The Urban Experience*, E. & F.N. Spon: London.

Fromm, E. (1976) *To Have or to Be?*, Harper & Row: New York.

Gieysztor, A. (1982) *Mitologia Slowian* (The Slav Mythology), Wydawnictwa Artystyczne i Filmowe: Warszawa.

Habermas, J. (1983) 'Na czym polega dzis kryzys? Problemy uprawomocnienia w poznym kapitalizmie' (orig. Was heisst heute Krise? Legitimationsprobleme im Spaetkapitalismus), trans. M. Lukasiewicz, in *Teoria i Praktyka: Rekonstrukcja Materializmu Historycznego* (orig. *Theorie und Praxis: Zur Rekonstruktion des Historischen Materialismus*), PIW: Warszawa.

Hall, E.T. (1987) *Bezglosny Jezyk* (orig. *The Silent Language*), trans. R. Zimand, PIW: Warszawa.

Hillier, B. (1989) 'The architecture of the urban object', *Ekistics*, no. 334/335, pp.5–21.

Jankowski, D. (1977) *Dom Kultury* (Cultural House) Centralny Osrodek Metodyki Upowszechniania Kultury: Warszawa.

Kapleau, Ph. (1988) *Trzy Filary Zen* (orig. *The Three Pillars of Zen*), trans. J. Dobrowolski, Wydawnictwo Pusty Oblok: Warszawa.

Kloskowska, A. (1980) *Kultura Masowa* (Mass Culture), PWN: Warszawa.

Krezel, E. (1983) 'Szlaki pielgrzymie w diecezji tarnowskiej' (The roads of pilgrimages in Tarnow diocese), in *Tarnoviensia Studia Theologica*, vol. IX, Instytut Teologiczny: Tarnow.

Kwiatkowska, A. (1994) 'Humanizacja blokowisk: Osiedle Zakrzow i Gaj' (Humanization of block complexes: Zakrzow and Gaj housing estates), unpublished report, Faculty of Architecture, Wroclaw University of Technology: Wroclaw.

Kwiatkowska, A. (1996) 'Interactive housing design', ENHR Conference in Denmark, unpublished paper, W 12, Danish Building Research Institute: Horsholm.

Kwiatkowska, A. (1997) 'Informative-interactive design theory of software age', in *Environment-Behaviour Studies for the 21st Century*, MERA Conference Proceedings, Department of Architecture, University of Tokyo, pp. 405–8: Tokyo.

Marx, K. (1951) *Kapital*, Instytut Marksizmu, vol. 1: Warszawa.

McLuhan, M. (1964) *Understanding Media: The Extensions of Man*, Signet Books, The New American Library of Canada Ltd: New York, Toronto.

Orzeszek-Gajewska, B. (1984) *Ksztaltowanie Terenow Zieleni w Miastach* (Shaping the Green Area in the City), PWN: Warszawa.

Parkes, D. and Thrift, N. (1980) *Times, Spaces and Places: A Chronogeographic Perspective*, John Wiley: New York.

Piatkowska, K. (1984) 'Zielen i wypoczynek' (Greenery and recreation), in M. Nowakowski (ed.) *Ksztaltowanie Sieci Uslug* (Shaping the Social Service Network), Warszawa: PWN.

Pietrasinski, Z. (1965) *Praktyczna Psychologia Pracy* (The Practical Psychology of Work), Wiedza Powszechna: Warszawa.

Radecki, A. (1990) *Piesza Pielgrzymka Wroclawska* (Wroclaw Pilgrimage), Wydawnictwo Diecezji Rzymsko-Katolickiej: Mokrzeszow.

Rojek, C. (1995) *Decentring Leisure*, Sage: London.

Rykwert, J. (1982) *The Necessity of Artifice*, Academy Editions: London.

Sicinski, A. (1977) 'Problemy przemian stylu zycia w Polsce' (Problems of life style changes in Poland), in J. Szczepanski (ed.) *Badania nad Wzorami Konsumpcji* (Research Studies of the Consumption Patterns), Instytut Filozofii i Socjologii PAN: Warszawa, pp. 281–364.

Statistical Yearbook 1992 (Rocznik Statystyczny), Biuro Statystyczne, GUS: Warszawa.

Stepien, M., Wilkon, A. (eds) (1989) *Historia Literatury Polskiej w Zarysie* (The History of Polish Literature), PWN: Warszawa.

Titman, M. (1995) 'Zip, zap, zoom: A–Z of cyberspace', *Architectural Design*, no. 118, pp. 48–51.

Wallis, A. (1993) 'Assimilation and accommodation of a housing innovation: a case study approach of the house trailer', in E.G. Arias (ed.) *The Meaning and Use of Housing*, Avebury: Aldershot, Brookfield, Sydney, pp. 425–41.

White, L.A. (1959) *The Evolution of Culture. The Development of Civilization to the Fall of Rome*, McGraw-Hill: New York.

Zaluski, Z. (1977) *Siedem Polskich Grzechow Glownych* (The Seven Polish Deadly Sins), Czytelnik: Warszawa.

Zygulski, K. (1981) *Swieto i Kultura* (Holidays and Culture), Instytut Wydawniczy Zwiazkow Zawodowych: Warszawa.

10

TOURISM AND SACRED LANDSCAPES

Richard Tresidder

Introduction

This chapter is concerned with the creation of a framework that attempts to understand our motivations as tourists, and why as tourists some landscapes figure above others, exemplified by National Parks. Many people wander no farther than two hundred metres from the car; buy an ice cream and then return to an urban home, yet feel as though they have experienced and consumed something unique, something magical. Certain touristic landscapes are understood as something extraordinary, special, perhaps even sacred. Holidays and certain landscapes can be used as temporal markers of lives and existence. There is no attempt here to provide a definition of a sacred landscape, as these are defined through reflexive practice. This chapter considers the way in which individual leisure landscapes may be elevated to a level that enables the experience to be defined as 'sacred'.

The tourism industry, tourism brochures and the hegemonic discourses of landscape provide one of the mechanisms in which we can explore the 'Self'. As Chris Rojek (1997: 53) asserts 'travel experience involves mobility through an internal landscape which is sculpted by personal experiences and cultural influences as well as a journey through space'. Tourism is a journey concerned with the search for dreams, and the creation of a touristic geography which is a correlation of personal and societal influences.

An investigation of the social and cultural processes of tourism consumption, has provided an insight into the motivations of tourists and the subsequent marketing of landscape by the industry. Such areas as the Lake or Peak Districts fulfil an important role in the process. Their representation of heritage, wilderness and rurality create a modern myth. We can see the representations of natural history and heritage, of rural-based economies, of ancient landscapes and the generally undeveloped nature of established tourism sites as the antithesis of post-industrial society, it is the creation of a myth of a country which possesses enclaves of bucolic, almost backward, insularity. We can see many tourism sites as a refuge from modernity in which we can find the organic, the primitive, the original and the expressive. These elements enable landscape to become defined as 'sacred'.

Sacred time and sacred places

To talk of the 'sacred' is to stimulate strong feelings in many of us, all of us define and construct notions of the sacred in a different way. This chapter does not attempt to destroy the semantic relationship of the sacred with traditional definitions of religion, rather, that in contemporary society tourism has become another form of religion, a means by which to justify our own existence. We can see the sacred as a temporal division in which we divide our lives into two distinct periods, that of the sacred and of the profane. The profane can be defined as that time we spend undertaking the daily activities to exist and live; these include working, shopping for food and housework. We can define the sacred as the time in which we are freed from the profane activities of existence; for Rojek (1993: 1) there are centres 'where we can shrug off the chained obligations of daily life and take a leap into "freedom" '. It is this freedom which Rojek has found in 'Greenfield Village' which marks the difference between the two categories. The sacred provides an outlet in which we can explore our reflexive self, to escape from the pressures of the profane, and to justify our existence through a complex celebration of both our individuality, and our society. The search for freedom is not dictated by the provision of sacred time and places, but through our own cognitive processes. For Urry:

> Tourism results from a basic binary division between the ordinary/everyday and the extraordinary. Tourist experiences involve some aspect or element which induces pleasurable experiences which are, by comparison with the everyday, out of the ordinary.
>
> (1990: 11)

The use of imagery by the Welsh Tourist Board provides an example. It has used the division between the sacred and profane (unwittingly or not) in a recent television advertising campaign. The campaign has used the differentiation between time and space to promote the Welsh countryside. There are two distinctions: the first is of space; they show the main picture in colour and use different landscapes to illustrate the beauty, wildness and freedom of the deserted Welsh mountains. There is an absence of people, houses, cars and the profane pressures of modern life. Second, to reinforce the difference between the Welsh landscape and urban landscape they use a small insert at the top of the screen. This shows pictures of crowded trains and commuters, the hustle and bustles of urban life, its profanity highlighted by the use of black and white film. There is a clear opposition between the two landscapes and each define each other, just as they define the differentiation between the sacred and the profane. Creating a temporal difference between the two landscapes further reinforces this differentiation. The grey urban picture has 'A.M'. printed beneath the picture, while the main colourful picture has 'P.M'. placed within the shot. This use of spatial and temporal images not only meets the desires of the tourist to consume sacred landscapes, but also creates the desire to escape from the profane elements of our lives.

This temporal and spatial differentiation is a constant theme throughout tourism marketing. In an analysis of eighteen Irish tourism publications, a constant theme runs throughout the publications; they support the notion of separation and distancing from the realities of everyday life. The recurrence of images falls into several categories:

- *Townscapes* show either deserted townscapes consisting of colourful houses or shops. They dress the human figures within the pictures in agricultural wear and they are driving a horse and cart or riding a bicycle.
- *Landscapes* are usually deserted wildernesses representing the wildness of the Irish landscape. If they include people within the image, they are isolated couples or a young family placed within the deserted landscape.
- *Heritagescapes* reinforce the notion of a mythical place in which we can search for roots and authenticity.
- *Workscapes*. All working scenes focus on the agrarian nature of the country; images of harvesting consist of men and women using pitchforks and other manual forms of gathering crops; similarly sheep and cows are placed within the deserted landscapes. All the images exclude the industrialisation of agriculture, again taking us back to a less hurried existence before the mechanisation of transport and industry.

These semiotic representations of place and people achieve the separation and differentiation of landscape from the realities of our profane lives. The differentiation provides the distinction and elevation of landscape into the realm of the sacred, a landscape where we can find a space to explore our reflexive and reflective selves.

However, there has to be a balance between the two divisions. If we have too much profane time we can no longer effectively function as an individual, just as if we possess too much sacred time we become bored and lack balance (Graburn 1977). Tourism has become one of the essential elements of everyday life, a means to escape the horrors related to the pressures of contemporary life, strong moral feelings attached to both work and play. Tourism becomes a justified escape from routine. Tourism takes on certain sacred qualities. In today's secularised society tourism provides one of the major forms of escape and, in consequence for some, the justification for living. The Protestant work ethic stated that if you worked hard, and saved hard, you would reap the benefits in the next life; in contemporary society you do not have to wait, tourism is the justification for work. Berlyn suggests that human life has to maintain a preferred level of arousal in 'artificial sources of stimulation . . . to make up for the shortcomings of their environment' (quoted in Stirrat 1984: 36). We can examine tourism against the ordinary workday life. Thus, events and institutions are the markers of the passage of natural and social time and ultimately become the definers of life itself.

Tourism at its most basic level is the production and consumption of dreams. Tourism becomes elevated to an almost metaphysical plane, although Cohen (1988), Lash and Urry (1994) disagree with the status of tourism in contemporary life; for Cohen tourism is an inauthentic activity. The processes of self discovery, escape and rejuvenation levitate tourism from the level of the profane into the sphere of the sacred. Religion has been argued practically as the human response to those questions that ultimately concern us. (Durkheim 1912, Weber 1978). Humans appear to be innately religious as we face the existential problems of modernity, ageing, disease and death. The implication of these two models is that tourism provides a release from the pressures of everyday life and vindicates our own existence; we are released from our normal social constraints in which we can explore our own identities and place in society.

Tourism as a form of ritual behaviour removes us from our everyday experiences, while simultaneously releasing us from our normal social constraints. Tourism becomes ritualised through, both the physical routines of tourism, and the entry from one classification of time into another. Durkheim (1912) perceived that simple forms of religion were rituals performed to celebrate society, promoting social solidarity and stability. Krippendorf (1987) defines tourism as a form of 'social therapy', as he states: 'Tourism is social therapy, the valve that maintains the world in good running order!' (1987: xv–xvi). As such, tourism has a stabilising effect not only on the individual but on our entire society and its economy.

When we consider the choice of touristic landscapes, it is interesting to observe Eliade's theory of distinction between the sacred and the profane as a product of humanity's 'terror of history' (1954). Especially pertinent to the cultural complexity of post-modernity Eliade argues that the sacred is outside time and space, and just as his discourse denies time, it also denies space, the sacred forms a reality separate from what we live. 'To be truly at one with the sacred involves attaining an existence outside time and space, and thus to be truly sacred, religious virtuosi must attempt to live outside society' (Stirrat 1984: 203). Tourism provides the opportunity to live outside both time and space for a limited period, whenever society agrees to release us the sacred becomes an attainable entity available to us all. If we accept the notion of the sacred, as either a temporal division, or a celebration of society, we have to accept that there has to be a place in which we can consume this sacred time, by definition it has to be 'different' or 'special'. National Parks enter the realm of the sacred, their importance learnt through schooling and television series e.g. 'Peak Practice' and 'All Creatures Great and Small', thus an element of our national consciousness (Urry 1995: 193–210). Their sacred existence is reinforced by the artefacts that represent the passage from the profane into the sacred: the Peak District's 'Mill Stone', Dartmoor's 'Lumps of Granite'. Whilst some UK National Parks abut cities and thus the profane, their location within the geographical boundary of the Park must classify the landscape as part of a sacred landscape. Yet the individual's classification is cognitive and reflexive. For some the countryside is firmly in the boundary of the profane.

The sacred in a secular world

Although we categorise certain landscapes as possessing sacred qualities, there is still a degree of resistance to using the phrase sacred in a secular context. The main resistance to the definition of sacred sites is the fact that it does not meet the traditional definitions of religion. However, what becomes labelled and adopted as sacred by society does not have to pertain to religion. Moore and Mayerhoff (quoted in Graburn 1977) argue that what is held to be sacred by society, such as the fundamental structure of beliefs about the world, may not be religious. Instead they may be felt to be essential to the continued national and individual constitution and thus, capable of stimulating strong emotions. They provide the example of Lenin's Tomb in Russia; although we can classify it as a touristic site, it is considered a sacred place in the context of an atheistic communist state, it provides a necessary focal point upon which to focus devotion and conceptions of being, a social marker which comes to represent struggles and the state itself. The distinctive social, cultural and historical significations of defined or adopted sacred sites, behaves as a depository for all the emotions of the alienated modern in search of meaning and the escape from anomie. Wedel (quoted in Graburn 1986) in her study of package tours of American tourists to Yellowstone Park, reports that many of the busloads of tourists spontaneously sang 'The Star Spangled Banner' or 'America the Beautiful', while ordinary tourists started to cry at the sheer magnificence of the landscape. In examining Elvis Presley's 'Graceland', Chris Rojek (1993) observed that:

> Hundreds (of tourists) linger on until the dawn breaks on the actual anniversary of his death, lighting a succession of candles. Throughout

the year, tour operators present visitors with the Elvis experience. Tourists are invited to walk where he walked, sit where he sat, see what he saw.

(1993: 143)

This elevation of sites to the sacred may become endemic to humanity. Eliade proposes that 'Every inhabited region has what may be called a "Centre", that is to say, a place sacred above all'. (1954: 39). If we consider that we possess a Centre within ourselves, a depository for all our culture, experiences and expressions of self, and rarely expressed, leisure and tourism provide an outlet for the projection of this onto the social. The desire of the modern tourist to search for these Centres, has been aided by the democratisation of travel and the perpetuation of touristic images of landscape in the media. This has enabled the tourism industry to provide unlimited access to places of escape with a constructed representation of history, culture and society. These representations gratify our cultural and aesthetic definitions of what is the sacred, through the provision of liminal spaces in which we can express the 'self'.

The metaphorical consumption of landscape

The physical existence of sacred landscapes is as important as the actual consumption of sacred places; we are safe in the knowledge that they exist, thus we always have the possibility of escape. This process of therapy or cleansing does not start with the physical consumption of landscape, but begins much earlier with the cognitive or metaphorical consumption. A package of some sorts surrounds everything purchased. Tourism is no different: just as other products are examined visually before purchasing them, similarly the packaging of tourism is examined through the brochure, the guidebook or the television programme. It is this packaging which fuels and meets our desires and hopes. Post-industrial commodity production has been accompanied by commodity packaging, the packaging has become a commodity in its own right, and we consume the images from the moment we first imagine or select the brochure. The tourism brochure catches the consumer's eye, it attracts attention, it promotes the notion of purity within its contents, it becomes a hygienically sealed product that is without history, an escape from contemporary society. As Susan Willis comments in her book, *A Primer for Daily Life* (1994) 'In the First world, the package is the fetishized sign of the desire for purity, in the fullest sense, it is also a desire for security' (1994: 3). Packaging becomes a dimension of the commodity form itself. Thus, we can see packaging as a metaphor for the formal contradiction of the commodity; Willis further develops this notion, and states:

In *Capital*, Marx initiated the analysis of the entire system of the capitalist economic relationships with an account of the commodity form.
This is the nexus of capitalism as well as the means of understanding
contradiction. Where Marx began with the commodity, I would begin
to understand the commodity as it is metaphorically reiterated in its
packaging.

(1994: 3)

Although we view landscape and culture on television and in the media, it is
only a temptation to consume, there is still a barrier between the physical
consumption and the semiotic consumption, but it prolongs the whole process
of consumption. If we accept that tourism is a form of social therapy, the healing
process begins with the anticipation we feel when we enter back into our everyday
routines. The construction of tourism sites separates the consumer from the realisation of the product, but it heightens the anticipation felt, we imagine the
experience of the place as we already expand its identity in the fullness of
the brochure. Willis (1994) in her analysis of soap operas identifies 'waiting'
as the most salient feature of soap operas. In both the contexts of soap operas
and tourism, waiting becomes a masochistic pleasure and an aesthetic and elevates
the consumer beyond the mundane reality of everyday life. The waiting involved
in the tourism industry from the booking, to the waiting to board the aeroplane
all help elevate the experience to the level of the sacred.

However, no commodity or landscape can ever live up to the expectations of
the consumer, in commodity capitalism's 'use value' cannot be fully recognised,
but rather it haunts the fetishised manifestations in the objects we consume.
Waiting can only be rendered aesthetically pleasing to the alienated, isolated
modern 'homeless tourist': 'the consumer learns to associate pleasure with the
anticipation of use value simply because commodity culture does not offer use
value itself as appreciable or accessible' (Willis 1994: 5). The tourism cycle
supports and promotes this fetishising of waiting, life becomes measured by the
temporal distance between allocations of structured leisure provision, the choice
of destinations and the departure from the everyday. The anticipation of actual
consumption is as important as the physical consumption of place in the overall
tourism process.

In search of the sacred

Why is the sacred landscape of our country, such as our National Parks, viewed
with such exhilaration? The consumption of sacred landscapes is seen to fulfil
some of our individual needs and desires, make sense of our existence. The landscape is not merely a physical entity, but is a social, reflexive, liminal space in

which we are freed from our normal social constraints, it is this freedom which defines the importance of such a segregated delineated space. Dean MacCannell (1976) viewed tourists as 'alienated moderns' in search of authenticity and experiences of the social. For MacCannell the mechanistic nature of work, the separation of work and home, and the other features of the advanced capitalist economic system contribute to a consciousness 'rootless' or 'homeless'. The consumption of tourism enables tourists to temporarily retrieve meaning from the scattered fragments of everyday life. Tourism appears to smooth the rough edges of everyday concrete social relations by the experience of being at leisure, of being free, of escaping.

With the onset of post-industrial society, the feelings of homelessness or rootlessness have become more apparent. The Post-Fordist economic system, and the hegemony of postmodernism as the cultural logic of late capitalism, actively supports fragmentation, semiotic commodification, and economic segregation; these elements of contemporary society have intensified feelings of discontent, rootlessness and ultimately anomie (Baudrillard 1988). The alienated modern caught in the uncertainty of postmodernity appears to be forever caught in the web of anomie. This socially experienced normlessness and meaninglessness associated with contemporary life does not offer an escape in ideology. Instead the 'sign culture' continually creates new 'wants' and 'desires' to the degree that 'wants' can never be achieved as the boundaries of desires are constantly changed and reconstructed, supported by the simultaneous creation of unattainable fashions, body images and social relationships. It is within this atmosphere that tourism may play a stabilising role for both society and the individual. For a limited period we are released from the social and economic pressure to consume. Although tourism is part of the consumerism process, it is constructed in a way to divert our attentions away from the horrors of everyday life. Tourism is an illusionary process that shrouds the processes of modernity. Sacred spaces act as a means of reference, their associations with nostalgia, heritage, community or the natural, allow us to find roots in a rootless world. According to Jameson (1985) people have lost their social and cultural foundations, they are lost both socially and spatially in the processes of de-differentiation of culture and society.

Sacred landscapes and the search for the authentic

The shift to a post-industrial economy and the break up of traditional working and community networks has heightened the desire for heritage-based visiting (Patrick Wright 1985). The search for authenticity takes on two distinct yet linked aspects: the search for authenticity of the self, the search for authentic experiences, together identified as the search for individual and collective meaning. Sacred landscapes can be seen as liminal spaces constructed by society, which provide us with the apparatus to escape from the very society that has created it. The liminal nature of tourism sites encourages a reflexivity of the self, which allows us to construct and deconstruct the landscape, figuring geographic knowledge, which consists of cognitive reflexive actions and individual refiguring of hegemonic discourses. However, to understand the touristic landscape we need to examine the underlying structure beneath any touristic landscape, as Tom Selwyn comments:

> beneath the surface structures of the various tourist sites, experiences, images and myths . . . , there remains a clearly identifiable sub-structure of concern in the tourist imagination with traditional-looking themes which seem at once modern and pre-modern and to which the term 'authentic' seems all too applicable, namely the nature of the social and of the self.
>
> (1996: 20)

The search for the authentic may then be considered as a search for moorings in a shifting world, and is concerned with the semiotic and physical consumption of the myths of tourism sites and their history. By consuming these, it enables us to use them as an oppositional key that acts as a guide for our own identity and place in society. The creation of sacred landscapes is to create a container of culture, society, history, and most significantly, space of identity that is highly abstracted but in practice also enacted. There is also a risk:

> Place-identity, in this collage of superimposed spatial images that implode in upon us, becomes an important issue, because everyone occupies a space of individuation, and how we individuate ourselves shapes identity. Furthermore, if no one 'knows their place' in this shifting collage world, then how can a secure social order be fashioned or sustained?
>
> (Harvey 1993: 308)

The tourist representations of such places as Ireland and Wales are imbued with unspoilt almost pre-modern landscapes, they are localities which belong to a time before the rationalisation of land ownership and the subsequent disenfranchisement of people from landscape. (See Morphy 1995: 205–45 for discussion of the reclamation of landscape in Australia.) The tourism industry has given us a stage and the power to reassert our ability to define and reclaim certain landscapes; even though this process may be illusionary, the feeling of freedom cannot be taken away. The ability to escape or to define our own manifestations of sacred landscapes can be seen as a true expression of the power of the self. Such freedom from our everyday lives, assumptions of power and industry, becomes the antithesis of the internationalisation and globalisation processes, these constructed landscapes with their associated conceptions of freedom become places which are unique and special, and ultimately sacred. Jameson's view on the spatial dilemmas of postmodernity illustrates this process, as he states:

> Spatial peculiarities of post-modernism and expressions of a new and historically original dilemma, one that involves our insertion as individual subjects into a multidimensional set of radically discontinuous realities, whose frames range from the still surviving spaces of bourgeois private life all the way to the unimaginable decentring of global capitalism itself.
>
> (1985: 351)

If we are to consume sacred time we need to create or locate landscapes which are outside both time and space. The expression of authenticity supports the status of these places, even though they may be our own definitions of the authentic, they continue to provide a feeling of security and existence in this shifting world. Foucault comments on the construction and function of place and space in the fabrication of identity in the postmodern world,

> The construction of such places, the fashioning of some localised aesthetic image, allows the construction of some limited and limiting sense of identity in the midst of a collage of imploding spatialities. . . . Space is fundamental in any form of communal life; space is fundamental in any exercise of power.
>
> (1987: 253)

The creation of sacred landscapes illustrates the power of the individual to seize and redefine landscape from a constantly shifting world, which constantly appropriates and commodifies cultural landscapes. Although the creation of touristic images and sites cannot be separated from other larger questions, such as notions of power, there remains a political worth in the creation of a stable visual aesthetic which reflects certain notions of power. This creation of a hegemonic view of sacred landscapes helps to create a common view. However, there still exists within the power relations of the self and society, the ability to be reflexive and cognitively expressive. It is this freedom, that occurs within the liminality of sacred landscape, that encourages the process of social therapy and ultimately the definition of sacred places.

Conclusion

In conclusion, the creation of sacred landscapes is a reflection of our ability to be free-thinking reflexive individuals in which we release 'ourselves' from our normal social constraints. Although we may ascribe certain characteristics to landscape through our existing geographical knowledge there exists a cognitive element, which enables us to link our own personal memories and preferences to established geographical characteristics. The ability to achieve this personal redefinition of landscape through leisure activities elevates landscape and experience to the level of the sacred. We may use certain stereotypical notions of place to reinforce our recognition of sacred sites, but the way in which we interpret them is the important ingredient in the sacred experience of tourism.

Our desire for sacred time and places may be defined in terms of power relationships supported by marketing and the needs of post-industrial society. But we cannot deny the associations we feel with certain landscapes such as our National Parks or the touristic images of Ireland. As I stated earlier, tourism at its most simplistic level is concerned with the production and consumption of dreams, the notion of the sacred enables us to provide some form of a theoretical framework to support the needs and desires of tourists. I sometimes think that as academics we attempt to over theorise what is an important element of all our lives, and by doing this we dismiss the fundamental element of reflexivity and the analytical ability of individuals and tourists to define their own experiences and landscapes. It is this innate reflexivity which enables the spatial and temporal activities of tourism to be defined as sacred. As such it is unlikely that we can give a general definition of what constitutes a sacred experience or landscape: only the individual can do this.

References

Baudrillard, J. (1988) 'Consumer Society', in M. Poster (ed.) *Jean Baudrillard, Selected Writings*, Stanford University Press: Stanford.

Cohen, E. (1988) 'Authenticity and Commoditisation in Tourism', *Annals of Tourism Research* 15(3).

Durkheim, E. (1912) *The Elementary Forms of Religious Life*, George Allen and Unwin: London.

Eliade, M. (1954) *The Myth of the Eternal Return*, Princeton University Press: Princeton.

Foucault, M. (1987) *Language, Counter-memory, Practice: Selected Essays and Interviews*, ed. Donald Bouchard, Cornell University Press: Ithaca.

Graburn, N.H. (1977) 'Tourism: The Sacred Journey', in V. Smith (ed.) (1986) *Hosts and Guests: The Anthropology of Tourism*, ATR, Vol. 10: 530–63.

Harvey, D. (1993) 'From Space to Place and Back Again, Reflections on the Condition of Postmodernity', in J. Bird *et al.* (ed.) *Mapping the Futures: Local Cultures, Global Change*, Routledge: London.

Jameson, F. (1985) 'Postmodernism and Consumer Society', in H. Foster (ed.) *Postmodern Culture,* Pluto: London.

Krippendorf, J. (1987) *The Holiday Makers*, Heinemann: Oxford.

Lash, S. and Urry, J. (1994) *Economies of Signs and Space*, Sage: London.

MacCannell, D. (1976) *The Tourist: A New Theory of the Leisure Class*, Macmillan: London.

Morphy, H. (1995) 'Colonialism, History and the Construction of Place: The Politics of Landscape in Northern Australia', in B. Bender (ed.) *Landscape, Politics and Perspectives*, Berg: London.

Rojek, C. (1993) *Ways of Escape: Modern Transformations in Leisure and Travel*, Routledge: London.

Rojek, C. and Urry, J. (1997) *Touring Cultures: Transformations of Travel and Theory*, Routledge: London.

Selwyn, T. (1996) *The Tourist Image: Myths and Myth Making in Tourism*, Wiley: London.

Stirrat, R.L. (1984) 'Sacred Models', *Man* (N.S.) June 1984, 19: 199–215.

Urry, J. (1990) *The Tourist Gaze: Leisure and Travel in Contemporary Society*, Sage/TCS: London.

Urry, J.(1995) *Consuming Places*, Routledge: London.

Weber, M. (1978) *Selections in Translation*, ed. W.G. Runciman, Cambridge University Press: Cambridge.

Willis, S. (1994) *A Primer for Daily Life*, Routledge: London.

Wright, P. (1985) *On Living in an Old Country*, Verso: London.

11

DESIGN VERSUS LEISURE

Social implications of functionalist design approaches in urban private gardens of the twentieth century

Gert Groening and Uwe Schneider

Introduction

Around 1900 a special design approach for urban private gardens was introduced to Germany. It was the idea of 'the garden as an amplification of the house' and it brought landscape architects, architects and, last but not least, the users of urban private gardens into conflict. This moment brings into focus a debate concerning leisure practice, identity and the use of space. Although the design approach is known as part of a historical debate in landscape architecture, its preconditions as well as its implications for the design of gardens and for gardening have not yet been researched. The contradiction between a user-orientation which the architect claimed for his concept, and his interest to check virtually all aspects of design is of special interest. We attend to whether an urban private garden should become arranged entirely according to general functional and formal-aesthetic demands of a professional designer or be a space which reflects the individual experiences and projections of a garden owner. Hence follows the question of how to design; or rather, who shapes leisurely pursuit in a garden?

Although our focus is on elaborate gardens which surround the suburban country houses of a wealthy clientele, the same may be found in smaller gardens of suburban garden-cities, small-house settlements, and in allotment gardens. Individual appropriation of 'designed' private urban space may be even more hindered by generally accepted instructions and the design approach of bureaucratic assessors (Crouch and Ward 1997: 8–11, Groening 1974: 58–62, Groening 1984: 755–60, Groening and Wolschke-Bulmahn 1995:177–200 and 249–51, Poblotzki 1992: 249–62). The exemplary aesthetics of the country house garden for minor types of urban private gardens, e.g. for gardens in suburban small-house settlements, and for allotment gardens, were expressed by Karl Heicke, the editor of the professional German journal 'Die Gartenkunst' in a series of programmatic contributions in 1918 and 1919 (Groening and Schneider 1994: 447–54). In England, the aesthetics of larger arts and crafts country house gardens were proposed for a less wealthy clientele in the early twentieth century (Weaver 1912). Hence, the significance of this design approach reaches beyond the concerns of a small privileged layer of society, where the user of the garden usually is the owner of a one-family house (Herlyn and Herlyn 1976: 82–3).

Until the end of the nineteenth century only a small percentage of the city dwellers in Germany lived in one-family houses with private gardens. The best documented city in this respect is Leipzig, where in 1890 only 0.8 per cent of the inhabitants in Alt-Leipzig and 2.3 per cent of about 180,000 inhabitants in Neu-Leipzig lived in one- or two-family houses with private gardens (more than two-thirds also owned their house) (Hasse 1891:182). In those days to live in such a villa-like house was a class-privilege. Later the situation improved somewhat with the broader distribution of wealth. In the late twentieth century between 13 per cent (West Berlin 1987), 17 per cent (Hamburg 1993), and 19 per cent (Hannover 1992) in cities with more than 100,000 inhabitants own a one-family house with a private urban garden (Tessin 1994: 29–34).

If one adds other categories of gardens, such as tenants gardens, allotment gardens and atrium gardens, the general provision of gardens increases. By 1988 about half of the households in West Germany had a garden (any kind), numbers varying with community size. In communities with less than 5,000 inhabitants 78 per cent owned a garden, in large cities with more than 500,000 inhabitants the percentage was 26 only (Tessin 1994: 100). In spite of the social and economic differences private open space has a high social meaning.

Recent social science-based research on open space planning has developed the theoretical construct of an ideal-typical living room which consists of indoor and outdoor space for living (Groening 1972: 11–15). In it the house is viewed as the indoor living room and the garden is viewed as the outdoor living room. This construct allows deviations and restrictions to be defined for those cases where people who would want to use private open space but cannot do so for various reasons, e.g. since their open space is spatially separated from their private built space i.e. their home. Formal or aesthetic criteria which concern the artistic lay-out of these private open spaces are not touched by this instrumental approach (Groening 1972: 11–15). In garden literature the term has found widespread response and covers different formal and functional aspects. During the first two decades of this century many architects and landscape architects clearly had in mind a formal and architectonic design for such an outdoor living-room/garden which related closely to the design of the house. Later, as a concomitant of the emerging criticism of formal design practices, the term 'garden as an amplification of the home' was also used for gardens with less formal design. Consequently it loosened its ties with the original meaning (Bacher 1995). Here we especially refer to the term as it was used in the decades around the turn of the nineteenth century. What does this term imply for the use of a private urban garden as a place for leisure? Reference to contemporary statements will prove that this professional design approach with its orderly, spatial layout obstructed other dimensions of horticultural activity and therefore severely restricted gardens and gardening as a leisurely pursuit.

Personal documents, letters, diaries from lay people and a few selected gardens from lay people serve as examples of an interest in horticulture and garden design that reaches far beyond the more formal and stylistic concepts advocated by professionals (Hunt/Wolschke-Bulmahn 1993). The existential meaning of private garden culture may be characterised:

> Town gardens – indeed any places in which people gardened in town – are not just diminutive or undistinguished objects; they are the scenes and products of a network of social, physical, and symbolic orderings of private living space. They are microcosms of everyday life – rich and largely unexplored testimonies to the habitual orderings of house and garden. Above all, they are scenes of intense personalization.
>
> (Longstaff-Gowan 1993: 48)

Thus private gardens express a spiritual and physical appropriation of immediate surroundings. Because of the significance of individual horticultural activity we accent especially those approaches which relate to process-oriented, temporary and participatory design. The examples we cite for that originate from different periods of time, and relate to different cultural, social and economic situations. This may lead to results which may become generalised with respect to the meaning of lay horticultural activity.

Recent research on this subject has elicited some fundamental motives for gardening as a leisure activity which support our assumption about the differences between the value lay people attach to gardening and the value professionals attach to designed gardens.

The emergence of the idea of 'the garden as an amplification of the house'

The garden can only serve as a spatial amplification of the house when it is immediately connected to the house. This idea dates back to antiquity. The Greek pastas- and peristyle-house (Carroll-Spillecke 1989) and the Roman atrium-house inextricably connected built up and open space (Greek: pastas, peristyle, kepos; Roman: atrium, hortus). A formally and functionally comparable connection is valid for open space adjacent to built up space in various ways. The architect Reginald Blomfield advocated a close relationship for the design of the house and garden:

> The question at issue is a very simple one. Is the garden to be considered in relation to the house, and as an integral part of a design which depends for its success on the combined effect of house and garden; or is the house to be ignored in dealing with the garden?
>
> (1892: 1)

The German architect Muthesius developed a history of the English country house. He objected both to a romanticising view of nature which he saw in a landscape garden, and a representative interpretation which he saw in the old aristocratic garden. For him the sportive element needed more significance, i.e. in a garden the design of playgrounds and tennis-courts. The house terrace, and the light and air baths in a modern garden should encourage enjoyment of fresh air outdoors. Special meaning was given to the orchard and the vegetable garden.

> For the modern garden . . . the aspect of use is eminent . . . Modern man's sense of use will enforce the preference of a productive garden to an ornamental garden.
>
> (Muthesius 1907: xxvi)

Muthesius engaged himself in a garden programme which should serve the interests of an imagined average person. Such a normative idea of people neglected individual traits and reduced them to a few generic needs and properties. His goal was to create a new bourgeois aesthetic. His elitist position made him believe that individuals needed to be enlightened about the true culture of living and of gardening. The general lack of culture which he associated with the average citizen needed to be improved:

> It is true that today's public demand for art is most unclear, even most uncultivated and rawest.
>
> (ibid.: x)

In order to reach a higher standard of taste the public needed comprehensive instruction. A small number of specialists who had already recognised 'the unbearableness of the hitherto existing state of art' should implement it. Muthesius called this group of leaders, in matters of culture and taste, an aristocracy of mind, and thus confessed his elitist understanding of culture (Hubrich 1981: 94).

As supporters of the new movement an aristocracy of mind is emerging, who now consist not of aristocrats by birth but of the best bourgeois elements, who thus characterize the new and enlarged goal of the movement: to create contemporary bourgeois art.

(Ausstellungskatalog 1990: 32)

Muthesius' claim to create contemporary bourgeois art aimed at a unified style that would subject to it all life, as he believed had been the case in earlier periods of art (Muthesius 1903). Originally Muthesius had demanded that the design of the architect should follow the needs of the client. In reality, however, this claim became superimposed by his own idea of an exclusively regular-architectonic design. The programme for a garden as an amplification of the house had offered an analogy only to the functional areas of the house.

From the view of a professional designer Muthesius' concept for the design of the garden as an amplification of the house clarified formal and functional design preconditions. The goal was a meta-individual design programme which reflected average 'needs' of an imagined garden owner. It seemingly covered the basic options for garden use, that is: representation, economy, leisure. For leisure activities these gardens offered active and passive use. Locations for active use were the lawn-tennis-court, the playground with sandbox and swing for little children, and the athletic ground with athletic equipment for older children. Terraces and verandas meant passive leisure such as reading, relaxing, and small talk. Every now and then separated garden spaces for sun and light baths were provided. According to the programme and the size of these gardens, spatially, precisely defined areas were designated for leisurely use. Leisure thus became determined both spatially and as to the contents. It seems questionable if such strict spatial functionalisation of leisure and other garden activities can aptly reflect the wealth of garden-related user interests. Before the latter aspect will be discussed we will show how Muthesius' contemporaries reacted to his attempt to aestheticise, functionalise, and spatialise garden design.

Aesthetification and design control

In the late 1970s the art historian Werner Hofmann named an interest in aesthetification, as represented by Muthesius, a 'Geschmacksdiktatur', a dictatorship in taste (1979: 55). In 1900 the Austrian architect Adolf Loos had already satirised such tendencies in architecture in a story about the poor rich man:

The architect wanted to be friendly with him. He had thought about everything. For even the smallest box there was a place provided. The apartment was comfortable but it was very intellectually demanding.

153

Therefore in the first weeks the architect supervised living, so no mistake would creep in. The rich man made every effort. Nevertheless it happened that he laid aside a book and put it on a shelf which had been made for newspapers. Or he dropped the ashes from his cigar in the hole of the table which was supposed to receive the candlestick. When one picked up an object there was no end to the guessing and searching for the proper place and sometimes the architect had to unroll the blue prints for details in order to rediscover the location for a box of matches.

(ibid.)

In such a house design a user was simply felt to be a disturbance. The situation for a garden was judged to be similar in England. In 1899 a contribution to the journal *Country Life* criticised this attitude:

The ambition of the architect of to-day – in many ways a noble ambition – knows no limits. He will sometimes prescribe for you the furniture which must be placed in the rooms of his designing, the plan and outline of the garden on to which his windows look, and the very plants and shrubs which must be grown in the various parts of the garden.

(Anonymous 1899: 272)

The professional superiority of the architect or gardener tended to restrict the implementation of the garden owner's own personal ideas. In memory to the garden of his parents the architect Charles Paget Wade (1883–1956) documented such a case:

But he [the gardener] became such an autocrat, it ceased to be our garden any more, it became his garden in which we were allowed to walk. If asked to move a plant, we were always told it was the wrong time of year, or the plant was too old or too young, or the moon not old enough, in any case it would die if moved. He ordered all the seeds and plants and put them where he wished, our only part was to pay.

(Ottewill 1989: 136–7)

Garden users felt confined by architectural design and professional dominance.

It was not so much the stylistic shape of the garden which was decisive, i.e. whether the design followed geometric-architectural or landscape forms. Also, here, external cultural influence which the garden owner may bring in plays a rôle. The participation of the garden owner in the design and in the implementation process for the garden seems more important. It appears that individual participation of the garden owner is crucial for the identification with the garden owned. The meaning of individual participation and appropriation for the garden owner necessitates to rethink Muthesius' somewhat narrow notion of a functional garden. Is there an alternative for those who want a garden and use it leisurely?

Alternative concepts for the functionalisation and aesthetification of gardens: 'suspensive work', self-determined activity and the processual appearance of gardens

Urban private gardens seem to elude a precise definition with respect to form and function. 'Every garden . . . should realise the owner's individual dream, and represent his own little bit of paradise' (Cartwright 1892: 218). These words address the subjective meaning of private open space. How then can the garden become a representation of paradise? 'It ceased to be our garden any more' was the complaint in another quotation. It barely concealed the disappointment of a garden owner who was declared a spectator, who could not realise and pursue his personal interest in garden design and gardening.

Voluntary work in a garden and an ever-changing design obviously seem to be an important aspect for many who want to enjoy a garden. M. d'Arbelay, an aristocrat, who once fled the French revolution of 1789, laboured intensively in his cottage garden. He was not particularly successful but enjoyed it very much, as reported by his wife to his father:

> His greatest passion is for transplanting. Everything we possess he moves from one end of the garden to another, to produce better effects.
>
> (Scott-James 1982: 28)

In sociological terms Juergen Habermas had labelled this function of work 'suspensive'. Habermas' point was that physical exhaustion was no longer a characteristic phenomenon in most professions in the 1950s and thus the need to regenerate physically during leisure hours had diminished. So instead of this regenerative function of leisure he found two other functions of leisure which he called compensatory and suspensive. The latter function of leisure Habermas described as:

> during leisure a work-behaviour is practised which suspends from the alienation, abstractness, and disproportionateness connected to professional work; this quasi-work shall bring back the freedom, the plainness, and steadiness of the claim for achievement which the other work denies. One is not ready to accept the denials, and does not want to compensate them only but wants to suspend in a precise sense: leisure

promises a fulfillment, which is true and has nothing of a surrogate satisfaction.

(1958: 224)

Recent sociological research points to the elementary meaning gardening has for a vast majority of garden owners (Tessin 1994: 132, Groening 1974: 51–5). One third of the garden owners in Germany have 'much fun' with garden work, and another third 'mostly fun' (Tessin op. cit.). Westmacott found similar results for the rural horticulture of African-Americans in the south of the USA (Alabama, Georgia and South Carolina). He met people who were deeply satisfied with hard garden work. Lucille Holley from Alabama told him:

> I was raised with the work and I enjoy it. Used to truck farm and it would call for before-day and after-night a lot of times but I was doing it for myself and I didn't mind.
>
> (Westmacott 1993: 91)

Some obviously perceived garden work as relaxing compensation for hard professional work. From hindsight Sarah Hull from Georgia let him know:

> We worked on the farm and then after we'd come in from picking cotton I'd go and hoe the flower yard, keep the grass out. I really loved working with flowers.
>
> (ibid.: 88)

This corresponds to the results of a survey in Germany in 1993 where 70 per cent of the garden owners said that for them work in the garden was not work in the real sense (Tessin 1994: 132).

Apart from the physical activity it is the organisation and the spatial arrangement of a garden which follow the ideas of the garden owner. Neither aesthetic preferences nor the resulting spatial ordering need to be different from those of a professional designer. 'Every garden should . . . realise the owner's individual dream, and represent his own little bit of paradise.' Such a paradise could be designed in an unpretentious way. A passage from a letter written in the late eighteenth century seems to confirm this. Here a lay person reports on her design activity in the garden:

> Nearly all afternoons I spend an hour in my new garden, I sow, plant and begin to feel that this self fostered child will make me happier than all I had here before. According to my present plan I will design a large oval bowling green, amidst which I will plant roses on a little hill in

front of the hall which will be our summer residence. It will be circled by a flower bed . . . I also plan an arbor full of climbing plants which make for the most beautiful of all arbors.

(Bender 1996: 34)

It seems to be the self-determined participation both in the physical and the disposing activities which is decisive. Only this participation allows for appropriation of the object.

Self-responsibility for the design of the garden is one of the most important motifs for garden owners. This is fairly independent from the kind and size of the garden. According to a 1993 survey in Germany 40 per cent of the garden owners saw the meaning of the garden in the fact that they themselves could realise their own design ideas (Tessin 1994: 134). In Hannover, Lower Saxony, in 1992 most garden owners had the feeling they could do more or less whatever they wanted in their garden. Differing results were due to the types of gardens, i.e. if the garden was a tenant's garden, a row-house garden, a single family house garden, an atrium garden, or an allotment garden (ibid.). The atrium garden, a garden type which is enclosed from all sides by built space yielded the best results. Here conflicts with neighbors were lowest, and privacy was highest. This seems to match a more general definition of 'home territory' by Lyman and Scott:

> Home territories are areas where the regular participants have a relative freedom of behavior and a sense of intimacy and control over the area.
>
> (Groening 1972: 11–13)

In many European countries house construction requirements are such that professional consulting seems inevitable for legal, technical and constructive reasons only. Also, most people tend to build a house where it will be difficult, every now and then, to remove walls, doors and windows. This seriously restricts lay activities in house building. For a garden the number of requirements is much lower. This widens the range for lay activities (Hunt/Wolschke-Bulmahn op. cit.: 6). The meaning of lay activities and ultimately the high percentage of women engaged in them has been noted:

> Gardening . . . is also one of the spheres where amateurs and professionals mix freely, and where the transition from one to the other is accomplished with relative ease. This is another reason why women have become so prominent as gardeners – they have not been inhibited by professional barriers and can slip into expertise without having had to make their way up a career ladder.
>
> (Penn 1993: 10)

157

One end of the range of physical space for lay gardening may be marked by a roof terrace or a balcony. The important thing is to understand how such heterogeneous garden programmes developed:

> Additions have been made from time to time . . . but no formal plan has been laid down, and no written direction given. Great Tangley gardens have thus a delightfully spontaneous character.
> (Anonymous 1898: 111 et seq.)

It appears that the design of such gardens was a succession of unplanned events in the course of time. Such self-responsible activity in a garden, which does not follow a pre-determined plan, seems to be valued as a special quality in a garden and of gardening over time. It is the most important motive for having a garden. There is a procedural character of so-called 'vernacular gardens':

> Vernacular gardens might also be those where maintenance and management were privileged over making, and where aesthetics was never a primary concern. Inasmuch as the vernacular garden belongs to a specific group or subgroup that changes within as the larger society around it changes, then vernacular gardens are to be understood – even more than elite examples – as a process.
> (Hunt/Wolschke-Bulmahn op. cit.: 3)

As a result a clear line between professional design activities and horticultural activities of lay people can be drawn:

> . . . 'vernacular gardens' may be taken as designating – in a preliminary way – gardens which did not come into being as a result of the powerful intervention on a site of some wealthy patron or some 'name' designer.
>
> (ibid.)

The one-sidedness of professional design standards may be based on elitist categories:

> I am reluctant to define or judge the vernacular gardens in strictly aesthetic terms . . . What we must look for in the vernacular garden are precisely those qualities which the expensive professional pleasure garden rejects. We must try to discover the personal involvement in the design, maintenance, and use of the vernacular garden, and its role as

image of the family and its dependence on the approval of the outside world, the passersby.

(Jackson 1993: 16–17)

The communication of horticultural knowledge rather relates to practical advice for plant growth than to design issues:

> This is well borne out in eighteenth and early nineteenth-century English gardening and horticultural literature, which was dominated by practical knowledge disseminated through manuals, dictionaries, 'daily assistants', remembrancers, journals, handbooks, pamphlets, pocket books, and above all 'kalendars', all of which were calculated not so much to make, as manage a garden.
>
> (Longstaff-Gowan 1993: 50)

> Why, then, presume to write a book about gardening? . . . One acquires one's opinions and prejudices, picks up a trick or two, learns to question supposedly expert judgements, reads, saves clippings, and is eventually overtaken by the desire to pass it all on.
>
> (Perényi 1983: viii)

A similar tendency exists among African-Americans in the south of the USA. Questioned about the spatial arrangement of her plants Perry Royal from Alabama said:

> I just put them when there is space, that's about it. They look like they designed theirselves when they get the space you know.
>
> (Westmacott 1993: 91)

As for the preference of a straight or a curved garden path, Shirley Hitchcock answered: 'Well, I don't know . . . I just go with it' (ibid.).

The impression of improvisation which many of the gardens surveyed by Westmacott evoke is highly visible because quite frequently second-hand material is used. So it is common e.g. to use cans, bottles, bricks and field stones as path borders. Used tyres, tin buckets, car wheels, and barrels serve as flower stands. Often the huts and sheds in the gardens are made from used building material, applied successively, and frequently stored on the lots thus adding to the messy image.

Following the evidence given, horticultural activities of lay people may thus become characterised. An owner of a house garden annually spends about

200 hours per year, and an allotment garden holder spends about 760 hours per year in his garden (Nohl 1991: 514–5). Gardens also serve as locations for the subsistence production of fruit and vegetables. They are private retreats for families as well as places where friends and relatives meet, and they are locations for play and sports. As indicated a garden visit is motivated to a considerable extent by the opportunity for physical work. Of special significance is the chance for self-responsible activity with respect to individual design options. Due to the various biographical, social, and economic preconditions, as well as via external cultural influence a wide range of designs is possible, which can be explained from a phenomenological point of view:

> All vernacular creations, whether gardens or buildings, are constituents, and therefore products, of our everyday life and world; they are artifacts of the cultural landscape formed through cultivation and imposed order – the product of practice, not theory. Every artifact, like every situation, is essentially local and general, private and public. A small town garden is both private and local inasmuch as it is a space rendered meaningful by being ordered and cultivated for private production; similarly it is public and general insofar as the exercise of creating such a meaningful domain entails drawing from shared, external cultural experiences.
>
> (Longstaff-Gowan 1993: 48)

Conclusion

The social value of such gardens is reflected in the vigour of their individual expression and their procedural character, which may become manifest in unusual aesthetics of the materials used. An assessment of the artefacts, based on traditional aesthetic categories for professional design, and for the artistic measure of 'high' products of culture, is inappropriate.

Since horticultural activities in private open spaces have comprehensive social meaning, to functionalise a garden as in the concept of the garden as an amplification of the house proves static, and one-dimensional. A garden designed in this way offers a narrowly defined range of leisure activities only. The chances for individual change and appropriation have become minimised respectively and have become declared undesirable for artistic-aesthetic reasons, yet these are the most important motives for horticultural activity. Practically and intellectually realised gardening by lay people thus becomes synonymous with consummated leisure.

Improvisation and change is crucial to the lay gardening project: procedures need to reflect these. Aesthetics which derive from such an understanding of

planning and design have been described by David Crouch in relation to allotment gardens in England:

> Each holder develops a particular way of working, and each plot becomes
> a small landscape in its own right, an unintended aesthetic.
>
> (1992: 1–7)

A precondition for such individual garden culture is a sceptical attitude towards
the superiority of the architect and designer in matters of taste and culture. The
functional quality of a design approach, similar to the one advocated by Muthesius,
tends to restrict the potential value of the garden as a leisurely space. Additionally,
the user-orientation tends to contradict the architect's interest to develop a
comprehensive design.

Plate 4 Conflict between gardeners and the owner of a garden gone mad

Source: Carl-Ludwig Paeschke, *Das grosse Buch der Gartenzwerge*. Frankfurt am Main: Eichborn,
1994, p. 92.

References

Anonymous (1898) 'Country homes – gardens old and new: Great Tangley Manor', *Country Life* 4: 109–12 and 144–7.

Anonymous (1899) 'Country homes – gardens old and new: Athelhampton Hall', *Country Life* 6: 272–8.

Ausstellungskatalog (1990) *Hermann Muthesius im Werkbund-Archiv*, Berlin: Werkbundarchiv e.V. und Museumspaedagogischer Dienst.

Bacher, B. (1995) 'Auf der Suche nach dem neuen Garten: Gartengestaltung zwischen 1919 und 1933/38 in Deutschland und Oesterreich', *Die Gartenkunst* 7(2): 282–90.

Bender, H. (1996) *Das Gartenbuch: Gedichte und Prosa,* Frankfurt am Main: Insel Verlag.

Blomfield, R. and F. Inigo Thomas (1892) *The Formal Garden in England*, London: Macmillan.

Carroll-Spillecke, M. (1989) *Kepos: Der antike griechische Garten. Wohnen in der klassischen Polis*, vol. 3. München: Deutscher Kunstverlag.

Cartwright, Julia (1892) 'Gardens', *The Portfolio* 23: 211–18.

Crouch, D. (1992) 'British allotments: landscapes of ordinary people', *Landscape* 31(3): 1–7.

Crouch, D. and C. Ward (1997) *The Allotment: Its Landscape and Culture*, Nottingham: Five Leaves Press.

Groening, G. (1972) 'Ueberlegungen zu Wohnraeumen im Freien und deren Ersatzformen', *Landschaft und Stadt* 3(1): 11–15.

Groening, G. (1974) 'Tendenzen im Kleingartenwesen dargestellt am Beispiel einer Grossstadt', *Landschaft und Stadt*, Beiheft 10, Stuttgart: Ulmer.

Groening, G. (1984) 'Gestaltung im Kleingarten', *Das Gartenamt* 33(11): 755–60.

Groening, G. and U. Schneider (1994) 'The allotment garden as a country house garden', in S.J. Neary, M.S. Symes, and F.E. Brown (eds), *The Urban Experience: A People-Environment Perspective*, London: E & F.N. Spon, 447–54.

Groening, G. and J. Wolschke-Bulmahn (1995) *Von Ackermann bis Ziegelhuette. Ein Jahrhundert Kleingartenkultur in Frankfurt am Main*, Studien zur Frankfurter Geschichte 36, Frankfurt am Main: Waldemar Kramer Verlag.

Habermas, J. (1958) 'Soziologische Notizen zum Verhaeltnis von Arbeit und Freizeit', in G. Funke (ed.), *Konkrete Vernunft,* Bonn, 219–30.

Hasse, E. (1891) '"Gaerten", Die Stadt Leipzig in hygienischer Beziehung', in *Versammlung des Deutschen Vereins fuer oeffentliche Gesundheitspflege,* Leipzig: Duncker & Humblot, 168–90.

Herlyn, I. and U. Herlyn (1976) *Wohnverhaeltnisse in der BRD*, Frankfurt am Main/New York: Campus.

Hofmann, W. (1979) *Gegenstimmen: Aufsaetze zur Kunst des 20. Jahrhunderts,* Frankfurt am Main: Suhrkamp.

Hubrich, H.-J. (1981) *Hermann Muthesius: Die Schriften zu Architektur, Kunstgewerbe, Industrie in der 'Neuen Bewegung'*, Berlin: Gebr. Mann.

Hunt, J. D. and J. Wolschke-Bulmahn (eds) (1993) *The Vernacular Garden: Dumbarton Oaks Colloquium on the History of Landscape Architecture 14,* Washington, D.C.: Dumbarton Oaks.

Jackson, J. B. (1993) 'The Past and Present of the Vernacular Garden', in J. D. Hunt and J. Wolschke-Bulmahn (eds), *The Vernacular Garden: Dumbarton Oaks Colloquium on the History of Landscape Architecture*, vol. 14, Washington, D.C.: Dumbarton Oaks, 11–17.

Longstaff-Gowan, T. (1993) 'Urban Gardening in England', in J. D. Hunt and J. Wolschke-Bulmahn (eds), *The Vernacular Garden: Dumbarton Oaks Colloquium on the History of Landscape Architecture*, vol. 14, Washington, D.C.: Dumbarton Oaks, 47–75.

Muthesius, H. (1902, 1903) *Stilarchitektur und Baukunst: Wandlungen der Architektur im 19. Jahrhundert und ihr heutiger Standpunkt*, Muehlheim-Ruhr: Schimmelpfeng cf. the English translation, in *Style-Architecture and Building-Art: Transformations of Architecture in the Nineteenth Century and its Present Condition. Introduction and translation by Stanford Anderson*, Chicago (1994).

Muthesius, H. (1907, 1910) *Landhaus und Garten: Beispiele neuzeitlicher Landhaeuser nebst Grundrissen, Innenraeumen und Gaerten*, München: F. Bruckmann.

Nohl, W. (1991) 'Ermittlung des Freizeit- und Erholungswerts staedtischer Freiraeume', *Das Gartenamt* 40(8): 510–17.

Ottewill, D. (1989) *The Edwardian Garden*, New Haven, London: Yale University Press.

Penn, H. (1993) *An Englishwoman's Garden*, London: BBC Books.

Perényi, E. (1983) *Green Thoughts: A Writer in the Garden*, New York: Vintage Books.

Poblotzki, U. (1992) *Menschenbilder in der Landespflege 1945–1970: Arbeiten zur sozialwissenschaftlich orientierten Freiraumplanung 13*, München: Minerva-Publikation.

Scott-James, A. (1982) *The Cottage Garden*, London: Penguin Books.

Tessin, W. (1994) *Der Traum vom Garten – ein planerischer Alptraum*, Frankfurt am Main: Peter Lang.

Weaver, L. (1912, 1913 1922) *Small Country Houses of Today*, London: Country Life.

Westmacott, R. (1993) 'The gardens of African-Americans in the rural South', in J. D. Hunt and J. Wolschke-Bulmahn (eds), *The Vernacular Garden: Dumbarton Oaks Colloquium on the History of Landscape Architecture*, vol. 14, Washington, D.C.: Dumbarton Oaks, 77–105.

12

CONSUMING PLEASURES

Food, leisure and the negotiation of sexual relations

Gill Valentine

Introduction

Food has come under scrutiny from anthropologists and sociologists (for example: Lévi-Strauss 1964; Douglas 1984; Murcott 1982, 1983a, 1983b, 1993; Mennell 1985; Mennell, Murcott and van Otterloo 1992; Lupton 1996; Charles and Kerr 1986, 1988; Beardsworth and Keil 1990, 1997; and James 1990, 1993) and more recently geographers too (for example: Cook 1995, Cook and Crang 1996, Bell and Valentine 1997, Valentine 1998a, 1998b, 1999). Within this body of work, research has emphasised the cultural relationships through which foodstuffs are constituted as social forms, the role of food in the reproduction of domestic social relationships and food as an example of the objectification of the relationship between different social groups and societies.

Despite the diverse nature of research in this area, there is a tendency, specifically, within the studies of contemporary Western societies, to emphasise the functional and symbolic aspects of food rather than to consider eating as a leisure practice and a bodily pleasure. Even the work on dining out – an obvious leisure practice – has tended to be concerned with the origins of restaurants as social institutions, the omni-presence or penetration of eating places within the urban landscape (Wood 1995) and the distinct cultural codes of bodily propriety and table manners that accompany the act of eating under the gaze of restaurant staff and other diners (Finkelstein 1989, Visser 1993). In this chapter, therefore, I don't just want to bring together some of the themes of previous research on food (including experiences of dining out and food within the home and discourses about body shape and sexual attractiveness), but I also want to refocus them by thinking about eating as a leisure practice, specifically the role it plays in the negotiation of sexual relationships in public and private spaces. It begins by considering food as a physical pleasure and the connections between food and sex; it then goes on to consider the importance that leisure practices involving food play in the initiation and management of sexual relationships; finally it explores the role of food in providing contingent knowledges about, and control of, other sexual bodies. While eating as a leisure activity can take many forms (for example, meals eaten at hotels, restaurants, public houses, fast food chains, cafés, take-aways, travel catering and catering at entertainment venues), this chapter will focus specifically on eating at restaurants and at home.

The chapter is based on the findings of a two-year project on food, place and identity funded by the Leverhulme Trust. This involved using multi-method qual-itative research techniques including food diaries, in-depth interviews, video diaries and participant observation to investigate contemporary patterns of domestic food consumption in a variety of Yorkshire households and institutions.

Food and sex

In medieval times when food was in short supply, eating for pleasure rather than out of necessity was not only a sign of wealth and privilege but also of gluttony, lust and avarice. The bodily pleasures of eating, which encompass not only taste, but also touch, smell and visual sensations, were and continue to be conceptu-alised as hedonistic pleasures – as indulgence. The privileging of the pleasures and vices, rather than the necessity, of eating within modern Western consumer societies has contributed to the development of popular food moralities in which the very definition of some foods as nutritious and therefore 'good for you' and others as 'bad for you' influences their desirability. James (1990) for example, points out the way that children are often rewarded for eating conceptually 'good' meat and vegetables by being given conceptually 'bad' puddings or sweets. The 'bad' food – the treat – is withheld if the 'good' is not eaten. In this way the very construction of particular foods as 'bad' contributes to our ex-perience of them as desirable, or in the words of a cream cake advertisement 'naughty but nice'.

The appetite for food has historically been linked to sexual appetite in Judeo-Christian thought. In medieval Europe, both were considered to represent animalistic desires and a lack of self discipline. 'To give into gluttony was consid-ered as opening the floodgates for other sins and vices, to allow the devil within the body' (Lupton 1996: 132). Both temptations of the flesh – gluttony and carnality – were eschewed by early Christians. Fasting was a way for women in particular to express their chastity and religious devotion – part of a quest for spiritual perfection. Even in modern secular societies the association between food and sex persists (Lupton 1996).

Writers, artists, film directors and advertising agencies have all used food as a metaphor for parts of the body. For example, in a book titled *Rude Foods*, the photographer David Thorpe, interweaves foods and women's body parts. Many different cultures use items of food – such as melons and salami – as nick-names for parts of the body or as terms of endearment, and focus on the mouth as a site of erotic desire (MacClancy 1992). The sensual pleasures of eating and sex share much in common. Biting, licking, sucking and chewing are all bedroom as well as dining-room practices. Indeed, particular foodstuffs such as oysters and truffles have a long history as aphrodisiacs (Bell and Valentine 1997).

Food advertisements often play upon these connections. The Häagen-Dazs ice cream campaigns (in which ice cream is smeared on naked bodies or fed by a partially clothed woman to an equally disrobed man) and the Cadbury's Flake advertisement (in which a woman longingly sucks the chocolate bar in the bath) both emphasise the erotic possibilities of these food products. Indeed, feminist critiques have argued that images such as these, and food-imaged toys such as the Strawberry Shortcake series of dolls and Tonka's Cupcakes which were popular in the US in the 1980s, elide food and women's bodies in such a way that not only are the women sexualised, but they are also commodified for men's sexual consumption (Gamman and Makinen 1994, Vardy 1996). Although Coward (1984) suggests that gourmet features and advertisements in women's magazines have also created a food pornography that seduces women in much the same way that conventional pornography aims to appeal to men. Indeed, Margaret Reynolds' (1990) collection of women's erotica features the recipes of the cookery writer Elizabeth David (Gamman and Makinen 1994).

Pleasure: food and relationships

Eating out is a growing form of consumption. The UK National Food Survey suggests that in 1990 each person consumed on average 195 meals outside the home while the consumer catering industry in the UK increased by 69 per cent in the six years between 1986 and 1992 (Payne and Payne 1993). Most of these meals are consumed not out of necessity but for pleasure. Eating with others represents an enjoyable way of spending leisure time with other people. Dining out is an escape from the mundane routines of cooking and washing up. It provides an opportunity to enjoy 'luxury' rather than everyday foods; and a performative experience in which dressing up and the spectacle or sense of atmosphere and occasion can be as much a part of the meal as the food itself. In particular, leisure practices involving food play an important part in the initi-ation and negotiation of sexual relationships, for example meals (both cooked by one partner for the other, or eaten out at restaurants) are often part of dating, celebrating or spicing up a relationship and a prelude to sex. Karen describes the role that eating out played in the beginnings of her relationship.

When I was first getting to know, it was interesting 'cause when I was first getting to know Colin there was just this whole period of this. Ahm, we had this whole – it was kind of like a courtship period . . . it was obvious from the very, very early point there was a mutual attraction. But it's like neither of us dreamed of saying anything. So it was this like, endless stream of going out for a meal. So ahm, the two places I have very fond memories of are Enoteca and the New Peking . . . And it was, it's interesting the thing about going out for a meal, ahm, I don't know what it is. It's like there's almost a safety in that. Neither of us was like, because then finally when, I won't tell you the story of what happened, but it was quite amusing, but I mean it ended up Colin asked me round for dinner and that was it really. It's really odd. But it was like we'd had all these meals – but that was kind of safe, you know. It was like – I suppose that for me was part of the thing about a meal at somebody's house being far more intimate.

As Karen described, restaurants and cafés represent a halfway house between the communal environment of a 'public' space, such as a pub or bar and the intimacy of a home. These therefore offer women the opportunity to meet unknown men in a 'safe' yet fairly 'private' and at the same time 'romantic' environment. These twin characteristics of 'privacy' and 'romance' also mean that restaurants are important sites for the playing out of adulterous affairs (Cline 1990). Cline quotes Robin, a secretary, who describes her experience of using restaurants as a meeting place:

When I started the affair with Tony, who was married to a girl in Dick's office, food assumed a great significance in our lives. It was almost more of a thrill than sex! We'd phone each other at the beginning of the week, pick a restaurant about twenty miles away, then drive separately to the rendezvous, parking in different streets about quarter of a mile from what we called base food camp. I'd get there first, and order avocado with prawns – always the same starter. I used to dip my fingers in the garlic dressing, and pour it slowly over the prawns, thinking about him, and imagining what we'd do. Then I'd rub the dressing over the avocado and wait for the phone call. He phoned the restaurant after he'd parked his car to make sure no one had spotted us. If I said it was all clear he'd arrive just as the waiter brought the second avocado.

(Cline 1990: 61–2)

Certain types of restaurants are produced as very heterosexual spaces, the erotic connotations of a shared meal being accentuated, for example, by the use of low lights or candles, and romantic music. Indeed, the performative nature of food service work means that some outlets have a policy of selecting waiting staff on the basis of their personalities. Zukin (1995: 154) argues that 'Waiters are less important than chefs in creating restaurant food. They are no less significant, however, in creating the experience of dining out.' In particular, flirtatious waiters are common in Italian restaurants, playing upon or acting up the stereotype of Latin lovers. The gendered codes of behaviour enacted by waiting staff also tend to articulate the assymetricality of hegemonic heterosexual relationships as Anna describes:

> The one thing which really does irritate me is sexism on behalf of waiters. I mean so many times when I was going out somewhere with Chris, I mean I know more than he does about wine, so I would order the wine – they'd pour it first for him to taste it and I would have to grab it and taste it. And then with the bill, I'd get my credit card out and they'll usually bring back the credit card slip to him, and that used to really bug me.

The heterosexuality of some restaurants – particularly in more provincial towns and cities – means that lesbians and gay men can often be made to feel uncomfortable or excluded in these environments. As Kate describes:

> I'm very conscious ahm, at this point in my life [she is in a lesbian relationship having previously been in a heterosexual relationship] of the kind of thing of eating out with your lover. And ahm sad, I don't feel that you can be well, I don't know, in terms of two women going out for a meal I think it's harder to actually be romantic in a public restaurant . . . in terms of having a romantic meal at a restaurant I still don't find it easy with another woman . . . you know really stupid things, I can remember going out . . . and I thought it was really odd that like they were going round with my – there were candles on the table and as people sat down they were lighting the candles as they sat down, and they sat us down and they didn't light the candle. And I was then trying to work out, did they do it because its perceived as being romantic and intimate and so two women – you don't do that then – or did they just forget?

Increasingly the growth of lesbian and gay eateries, particularly in the gay villages of Manchester and Soho, London, and the policy of some mainstream restaurant chains to recruit suitably camp gay men to provide what Crang (1994: 693) describes as 'a bit of homosexual frisson' are changing the sexuality of some eateries.

When Gillian, a senior civil servant, began a lesbian relationship for the first time, her partner's concerns about the 'heterosexuality' of restaurant environments, and fears that the 'intimacy' and 'privacy' of the restaurant (which Robin quoted above, as a heterosexual woman so enjoys) would actually signal the nature of their relationship to other people, meant that they avoided eating out. Gillian explains:

> Meals with Frank were routine, public, before or after the pub, part of an everyday steady relationship that people approved of. Then suddenly I had my first homosexual relationship with somebody who couldn't bear to be known about, so I went underground. There was no way we could eat out together. My social life completely changed. So did my feelings about food. I stopped entertaining. She cooked for me and I cooked for her, quietly, with great love. Cooking for her and with her was a real joy, a mark of affection. I had never found that before . . . I remember the foods I ate with Edith because of the intensity of feeling.
>
> (Cline 1990: 62–3)

Sharing a meal is a pleasurable leisure activity but it is also a performative act through which diners both construct a narrative of their individual identities and are able to develop contingent knowledges about each other. For example, Bourdieu (1984) famously described the knowledges that we have about which foods to choose, which cutlery to use and how to look after our bodies as 'cultural capital', arguing that practices of the self, such as eating, betray people's origins or *habitus* (internalised form of class conditioning). While Finkelstein (1989) has highlighted the fact that dining out is a mannered activity disciplined by customs in pre-figured ways of behaving. Food and meals, like other forms of consumption, also provide currency for everyday discourse and even a way for people to talk about their relationships. They are both the objects of conversation and facilitate conversation. Thus through the performative act of sharing a meal individuals can articulate their identities and competence in public culture and so assess their compatibility and even begin to define shared narratives of identity. As Juliet describes:

> The first time we ever went out I was really nervous 'cause I didn't really know him and I'd never been to the place we went to. But we just got on like a house on fire. The restaurant was really pompous and I did something stupid, and instead of looking down on me he just cracked up. We sat there for ages and ages after the meal having coffee and brandies and just talking. After that we never looked back.

'Eating together is held to signify togetherness' (Mennell *et al.* 1992: 115) both in terms of wider social groups but also individuals. It is the closest field of inclusion. As Douglas and Isherwood (1979: 61) have argued: 'consumption, commensality and cohabitation are forms of sharing'. Cooking for someone else at home can therefore mark the beginnings of a transition from eating together as a leisure (and indeed pleasure) activity towards living together where food preparation can take on the mantle of a routinised chore. Although, it has been argued that heterosexual women can derive both pleasure and identity from providing meals which their partners enjoy and that learning what men like by trial and error is an important part of setting up home together (Murcott 1983b, Charles and Kerr 1986, 1988). Dorfman (1992) has traced the history of the message 'the way to a man's heart is through his stomach', demonstrating how food industries, advertisers, women's magazines and cookbooks have promoted the idea that women should spend their leisure time preparing desirable foods for men. The flip sides to this of course are that women's failure to provide good food is often used by men to justify domestic violence (Dobash and Dobash 1980); and that women often see food preparation for others, not as a leisure activity, but rather as a chore that prevents them enjoying leisure activities.

For each of us food is appropriated into our own personal economy of meaning. We all engage in a process of value creation through our various daily practices around food. Living together therefore requires the reconciliation of individual practices and values and the construction and assertion of a shared identity through the display and use of food. Like the foods we choose in restaurants, the meals we choose to prepare and consume at home are also constitutive of our understandings of ourselves and others – part of a process of distinction. The physical arrangements for food preparation and the way that meals are incorporated into our daily routines also establish (and defend) spatial and temporal boundaries between partners, so that the control of food, the use of food to articulate love or care for others and so on are all part of the continuous work of social reproduction that goes on within households.

The difficulties of creating shared food practices and identities are evident in Kate's lesbian relationship with Julia. Their relationship was cultivated over romantic meals in restaurants. But when Julia stayed at Kate's flat their different approaches to cooking and eating exposed differences in their identities which led to the break up of their relationship. There had never been any emphasis on food and methods of preparation in Kate's childhood. She had largely learnt to cook by osmosis, picking up different things from different partners. When she began to cook with Julia she became conscious that Julia was, what she termed a 'food snob'. While Kate's usual fruit was an orange, Julia preferred quinces; and while Kate chopped her vegetables randomly, Julia was quick to point out the proper way to prepare and cook different items of food. Julia's cultural (and indeed economic) capital intimidated Kate. Fearing Julia's disapproval she became nervous and uncomfortable about cooking in her own home. Tensions also arose at breakfast time because Julia put pressure on Kate — who never ate breakfast — to eat with her at 7 a.m. On the one hand Kate described this as 'nurturing' and recognised that it was one way that Julia articulated her love for Kate, but on the other hand she perceived it to be controlling or smothering behaviour. Although Kate had adapted her consumption practices and relocated her own narrative of identity in order to adopt a shared form of identification with her previous partners (including becoming a vegetarian when she was in a relationship with a non-meat eater; and adapting her eating habits to fit in with the dietary needs of a previous partner who was a coeliac), she was unable to reconstruct her narrative of the self to accommodate Julia's eating practices.

Food, leisure and the regulation of sexual bodies

Food does not only play an important part in the negotiation of sexual relationships, but also in the definition and shaping of sexually attractive bodies and is therefore implicated in many women's leisure practices. The body is an object of desire. At particular historical moments different body shapes and sizes have been considered sexually (and socially) desirable. Tracing the historical evolution of the ideal female body, Gordon (1990) points out that an anorexic body shape has become the idealised standard of female beauty in fashion and media culture. Bennett and Gurin (1982) dub this contemporary fetishisation of thinness the 'Century of the Svelte'. Popular culture (including advertisements, magazines and so on) has played a key role in contributing to this process of eliding slimness with health and sexual attractiveness in what Lupton (1996) has termed the 'sex, health, beauty triplex'. Contemporary women's magazines and tabloid newspapers often feature articles and stories about diets, exercises or leisure activities (such as aerobics) which implicitly suggest that a particular body shape is the key to a successful sexual relationship (Gamman and Makinen 1994). The pressures for women to produce the space of their bodies in a sexually desirable way is, some writers argue, one of the key factors that can trigger eating disorders such as anorexia and bulimia (these and other explanations are reviewed in McSween 1993).

Thus the way heterosexual women feel about and manage the space of their own bodies (e.g. through diet and leisure practices) is intimately connected with their awareness of the sexual gaze of other (usually male) bodies. Jackie,[1] a lone parent, describes the importance she thinks men place on the size and shape of women's bodies.

You're definitely not as attractive to men when you're fatter I don't care what people say men like thinner women, they don't like big fat women. It's news to me if you find anybody that does – let me know because I just don't know where these men are. I'm convinced over years of like watching because I'm one of those people when I'm sat in a pub if there's nobody to particularly talk to I like to watch what's happening around me and you can see it all the time you know. And when you do see a very big woman with a normal size man if you want to call him that you sort of think umm that's unusual that doesn't happen very often and then you sort of usually find out there's problems with the relationship anyway. So I think oh right men definitely prefer thinner women . . . I think there's – if they were all honest which in my opinion men aren't always that honest but. But I think if they were all honest they would have to say that a few rolls of fat on a woman are not particularly attractive.

It's a view endorsed by Colin who describes his reactions to overweight women:

I am conscious about in terms of who I'm, the people I'm attracted to. And overweight for me isn't attractive, and I do, I'm awful because I do put it down to food in terms of if someone's overweight I'm very judgmental, I form opinions when I don't know any facts in terms of whether it's health issues for that person and it isn't to do with food, do you know what I mean? Which is a dreadful thing to say, you're not gonna, oh God, um [trails off in embarrassment].

Not surprisingly therefore most of the heterosexual women interviewed admitted to having changed their leisure practices significantly at various stages of their lives in order to make themselves sexually desirable for a man in particular, or men in general. Either, by taking up enjoyable leisure activities such as aerobics, swimming, running and so on, or through denial, for example by curtailing other leisure activities, such as going out for a drink or for meals. This supports the findings of a study conducted by McKie *et al.* (1993) which found that while the women she interviewed resented having to discipline their bodies in order to conform to an ideal body image of slimness/health/beauty, at the same time they still pursued the ideal.

The pressure to practise self surveillance and to exercise self restraint under the watchful gaze of other sexual bodies does not stop for many women when they are in a sexual relationship. Rather David Morgan (1996: 132) has used the term 'bodily density' to describe the way close spatial proximity to a person over a period of time 'can result in knowledge, control and care of each others' bodies in numerous repeated and often unacknowledged ways'. Sexual relationships give us some claim to, or rights over, another person's body — particularly to look at it, touch it and comment upon it. In the eyes of a sexual partner our bodies are a spatially open location. It is in a sexual relationship that the space of our body, with all its unsightly bulges that we guiltily conceal from family members, friends and colleagues under our clothing, is exposed to the sexual gaze of our significant other. Not surprisingly, therefore, within sexual relationships bodies can become contested terrain. Several women described emotionally charged battles they have with their partners about their eating habits, leisure practices and the way they manage their bodyspace. Carol explains how her bodily boundaries are eroded:

> I mean there are times when I say 'Right I'm going to be good for the next month, cut down you know, I know I've put on weight' and then two days later I'll be eating something and Mike will say to me 'What are you eating that for I thought you were trying to lose some weight' and er it gets me mad, but it probably gets me mad more because I know that I've said that and I feel bad about it, rather than get mad with him, he's probably doing the right kind of thing . . . I've got no willpower, I'm exactly the opposite of Mike from that point of view. If he says he's going to do something like that, that's it he's doing it, whereas if I say something like that you know, by day two I've lost yeah I've, I've very little willpower.

The time-space constraints of household routines place important limitations on individuals' corporeal freedom. For women in particular the timing of household meals, school or office hours and children's leisure activities and bed times mean that it can be difficult for women to find any leisure time in the day when they can exercise. Even where household members have the time to indulge in leisure activities like swimming, jogging, or exercise classes, the stresses and exhaustion engendered by domestic or paid work can make going out for a meal or a drink a more desirable leisure activity than exercising. If one member does spend their leisure time taking part in sport or exercise it can have negative consequences for other household members, who may have to take responsibility for childcare, meal preparation and so on, while the other is out. Tasmin describes the lack of leisure time she has because of her domestic commitments and thus the lack of corporeal freedom that she has within the home.

> I've got good intentions but they never work. I mean I like to go swimming but it's like finding the time when I can go swimming when I haven't got the children, or I haven't got to go and pick children up

from nursery and things . . . just doesn't work . . . I would like to be
very much thinner, you know.

It is not only concerns about what others think that pressurises women to discipline their bodies. Rather, some women argued that it was often feelings of
discomfort with their own bodies – for example, a restricted sense of spatiality
as a result of tight clothing, the sensation of 'wobbling' flesh or clumsiness when
walking quickly and a general sense of heaviness and lack of mobility – which
motivated them to spend time dieting or exercising.

In particular, leisure activities based around food and drink are a source of
guilt for many women. As Bakhtin (1984: 281) has argued 'by taking food into
the body we take in the world'; we incorporate it into ourselves and it becomes
part of us. In this way, the process of eating and drinking (especially heavy meals
or drinking to excess) can be experienced as invasive, and contaminating. Kate
describes the sense of pollution and self disgust she experiences when she is not
self disciplined about what she eats:

> I am conscious I suppose of what it's [food] doing to me. I mean particu-
> larly at times of going out for meals and stuff where – it's not that you
> feel conscious, you think God I've put on a stone or something you
> just feel bloated, heavy and uncomfortable. That isn't enjoyment to me
> when you leave a restaurant and you just feel so, oh disgusted with
> yourself.

Yet, while eating can be experienced as contaminating or polluting, it can also
be an enjoyable activity as part of socialising, a hobby or a comforting sensation
that compensates for feelings of boredom, or emotional troubles. In this way the
bodily and social pleasures of eating as a leisure activity can temporarily outweigh
the social pressures women experience to limit their consumption. Thus while
all of the women interviewed feel pressure to pursue these ideal body shapes to
a greater or lesser extent (as described above), most could not be regarded as
having disciplined or 'docile bodies'. Rather, many women engage in a boom-
bust relationship towards their bodies. For periods they pursue an ideal shape
according to moral representations about sexual desirability or the disciplining
gaze of sexual partners in which they diet or change their leisure activities,
substituting eating out or going out for a drink for more ascetic leisure practices involving sport and exercise. But these episodes are often punctuated by
periods when the physical and social pleasures of eating or drinking take precedence. Other women adopt a fatalistic approach towards their bodies; they
argue that they cannot control its demands for or enjoyment of food; or claim
that the pleasure deprivation of cutting back on enjoyable activities such as eating

and drinking, and the self discipline necessary to pursue more ascetic leisure activities are not worth the bodily gains. As Wendy explains:

> I think I'm pretty sort of, I don't know what the word is – fatalistic or whatever, you can get run over by a bus tomorrow. You know, that sort of thing, so why, why make yourself miserable in the meantime, I mean if you enjoy eating and enjoy drinking and it's not affecting our lives too badly, then why stop what you enjoy doing or making yourself miserable by trying to change your lifestyle if that's not what you want.

Indeed, she goes on to promote a common belief that it is equally as bad to be too self controlled or to spend too much leisure time exercising as it is to be too self indulgent, rather she claims it is more important to be happy whatever your appearance. She also argues that there is a limit to the extent to which individuals can change their own bodily shape, judging it to be a matter of biological lottery. She despairs of her colleagues who often refuse to socialise after work over a drink or a meal because they are worried about their weight. As she explains:

> I'm quite big boned, well we all are really, so I know I'll never be slim however hard I try, nor will the kids. We're just not made like that. So I don't see the point in making us miserable for it. I mean people at work that are on diets and watching everything they eat, they seem so miserable, it doesn't seem worth it and they say 'No, I'm not going to come out with you that night because I won't be able to drink anyway [laughs]'.

Women's preoccupation with managing the space of their bodies is not shared by men according to Millman (1980). She argues that it is only gay men who worry about their weight, claiming that heterosexual men do not perceive that their body weight affects their sexual relationships, career or masculinity. More contemporary research, however, suggests that men too are increasingly the subject of discourses, evident in magazines, television advertisements which commodify the male body, and that they too are becoming more self conscious of the surveillant gaze of other sexual bodies. White and Gillet (1994: 32) suggest that:

> Whereas formerly (before the crisis of masculinity and the emergence of the male body as spectacle in consumer culture) it was women who watched themselves being looked at in order to gain control over how they would be treated by men, men have also turned themselves into

objects of vision. Watching their physical selves, and imagining them-selves being watched by others, is a consequence of the crucial importance of physical appearance to their sense of self-worth for both men and women.

Exploring the world of male body building, White and Gillet (1994) argue that a muscular physique is a sign of embodied masculine power and a form of phys-ical capital which can be converted into sexual capital. This discourse is also evident in UK men's 'lifestyle' magazines. *Men's Health*, for example regularly includes features about how to lose your belly or how to get a six-pack stomach (e.g. *Men's Health* March 1997: 10 nutrition lies debunked; bigger chests, no weights; high-speed, time-saving workout; *Men's Health* April 1997: The ultimate route to the six-pack stomach; your 10-minute home workout). While some of these magazine features implicitly make the connection between physique and sex, the January/February 1997 issue was more up front. In an article titled 'athletic sex: a training manual' readers were given a series of exercises (including stretching, weights and squats) to get their bodies in shape for a range of sexual positions. Like women's magazines, men's magazines commodify the male body – the emphasis is on enhancing particular body parts through purchasing products, following work out programmes or changing diet (Jackson 1997). Although, whether male readers are responding to the magazines' emphasis on the body beautiful is more doubtful. While men's magazines have a significant and growing circulation, men taking part in focus group discussions as part of Jackson's (1997) study of the magazines claimed that they were casual readers and did not to take the articles seriously; an approach to men's magazines which some of the participants contrasted with the way that they imagined their part-ners to read comparable women's magazines.

Despite the increased concern about men's bodies exhibited by men's mag-azines the social meaning of being overweight remains highly gendered. Schwartz (1986: 18) argues that while fat women are objects of derision, fat men are still associated with money, power and sex. She claims that 'Where fat men inspire or terrify, fat women draw the camphor of sympathy and disgust – sympathy, because they cannot help themselves; disgust because they are sexually ambiguous, emotionally sloppy.'

Certainly, a concern with bodily appearance and sexual attractiveness was not apparent among the majority of the men interviewed as part of this study. Only two expressed any concern about the need to diet or change their approach to leisure time by switching from eating and drinking to sport and exercise. Rather, while women's bodily experience appears to be largely structured through their sexual attractiveness, men's bodily experiences were primarily structured through the category of health and the functionality of their bodies (i.e. men recognised the need to stay slim in order to avoid ill-health, strokes, heart attacks etc.). In other words men conceive of their bodies in 'active' terms, whereas women are more concerned with their appearance – in other words with their bodies as 'passive' objects (Saltonstall 1993). This is also evident in the way men and women conceptualise the male body within sexual relationships. While women claimed to feel pressure to manage their bodies in order to maintain their sexual desirability for their male partners, men (and their partners) argued that they have a responsibility to maintain their body's health in order to live as long as their female partners. This reflects the different way that men and women's bodies are idealised and valued in contemporary societies. Gender differences were also apparent in the way men and women policed each other's bodies in a sexual relationship. While women complained that they resented the way men criticised their bodies, men were more relaxed about the surveillant gaze of their partners, joking about women 'nagging' them to eat healthily and give up sweets and biscuits. In this way women appear to internalise men's critiques of their bodies, yet men brush off reciprocal comments, turning them back on their partner's by accusing them of 'nagging'.

The evidence of this section therefore is that women's bodies in particular – and to a lesser extent men's too – are constituted within sexualised discourses, which emphasise the responsibility we each have to maintain the space of our bodies in a heterosexually desirable way, either through restricting hedonistic leisure activities such as eating and drinking, or through engaging in more ascetic leisure activities such as sport and exercise. The practice of sexual relationships in turn confers 'ownership' rights over the body of (an)other, with the result that our bodies, as Morgan (1996: 120) points out, can become an object of, or an extension of an other's desires or projects (again there are gender differences in the practices through which men and women attempt to regulate and shape each other's bodies in heterosexual relationships). Thus the space of our bodies can be shaped through the density of their relationships to the space of another particular body.

Conclusion

This chapter has focused on the importance of leisure practices involving food in initiating and negotiating sexual relationships and has considered the role food and eating play in providing contingent knowledges about, and control of, other sexual bodies. As Fisher (1943: 353) argues: 'It seems to me that our three basic needs for food, security and love, are so mixed and mingled and entwined that we cannot think of one without the other.'

In doing so, the chapter has exposed two main tensions. First, the ambiguity which arises when shared food consumption is a leisure practice and when it is a routine, mundane chore. Certainly, eating out is regarded as a leisure activity; yet cooking at home for a partner can oscillate between being a pleasurable leisure activity, an act of love as part of the moral economy of the household, and an everyday and unwelcome chore.

Second, the consumption of food produces a tension between the pleasurable and hedonistic dimensions of food as a leisure activity in which it articulates identity, bodily pleasures, sexual desire and so on (which are often seen as excessive or self indulgent), and discourses of self control in relation to food and drink, and the pursuit of more ascetic leisure practices such as sport and exercise, in order to maintain sexual attractiveness. Yet of course, self discipline through the denial of food or through leisure practices involving sport and exercise can also be pleasurable in that these practices can produce a sense of achievement of control over the body or pleasure from the way others admire a disciplined body; while hedonistic consumption or leisure practices – eating rich or heavy meals or drinking to excess – can also induce unpleasant feelings of guilt or shame. Thus in contemporary leisure and consumer culture there is, as Lupton (1996: 153) has argued: 'a continual dialectic between the pleasures of consumption and the ethic of asceticism as means of constructing the self: each would have no meaning without the other'.

Acknowledgements

I am very grateful to the Leverhulme Trust for funding the research (award F118AA) on which this paper is based. Many thanks go to Beth Longstaff for all her hard work and enthusiasm while employed as the research assistant on this project. I am grateful to David Crouch for his helpful comments on a first draft of this chapter. Some of the material included in this chapter was first published in *Environment & Planning D: Society and Space*.

Note

1 All the names of people referred to in this paper have been changed in order to maintain their anonymity and confidentiality.

References

Bakhtin, M. (1984) *Rabelais and His World*, Bloomington: Indiana University Press.
Beardsworth, A. and Keil, T. (1990) 'Putting the menu on the agenda', *Sociology* 24: 139–51.
Beardsworth, A. and Keil, T. (1997) *Sociology on the Menu*, London: Routledge.
Bell, D. and Valentine, G. (1997) *Consuming Geographies: we are where we eat*, London: Routledge.

Bennett, W. and Gurin, J. (1982) *The Dieter's Dilemma*, New York: Basic Books.

Bourdieu, P. (1984) *Distinction: a social critique of the judgement of taste*, London: Routledge.

Charles, N. and Kerr, M. (1986) 'Food for feminist thought', *Sociological Review* 34(1): 537–72.

Charles, N. and Kerr, M. (1988) *Women, Food and Families,* Manchester: Manchester University Press.

Cline, S. (1990) *Just Desserts*, London: Andre Deutsch.

Cook, I. (1995) 'Constructing the exotic: the case of tropical fruit', in J. Allen and C. Hamnett (eds) *A Shrinking World?*, Oxford: Open University Press.

Cook, I. and Crang, P. (1996) 'The world on a plate: culinary culture, displacement and geographical knowledges', *Journal of Material Culture* 1: 131–54.

Coward, R. (1984) *Female Sexuality: Women's Desire Today*, London: Paladin.

Crang, P. (1994) 'It's showtime in the workplace: geographies of display in a restaurant in south-east England', *Environment and Planning D: Society and Space* 12: 675–704.

Dobash, R. E. and Dobash, R. (1980) *Violence Against Wives*, London: Open Books.

Dorfman, C. (1992) 'The garden of eating: the carnal kitchen in contemporary American culture', *Feminist Issues* 12: 21–38.

Douglas, M. (ed.) (1984) *Food in the Social Order: studies of food and festivities in three American communities,* New York: Russell Sage Foundation.

Douglas, M. and Isherwood, B. (1979) *The World of Goods: towards an anthropology of consumption*, London: Routledge.

Finkelstein, J. (1989) *Dining Out: a sociology of modern manners*, Cambridge: Polity Press.

Fisher, M.F.K. (1943) 'The gastronomical me', in *The Art of Eating* (1976), London: Vintage.

Gamman, L. and Makinen, M. (1994) *Female Fetishism: a new look*, London: Lawrence & Wishart.

Gordon, R.A. (1990) *Anorexia and Bulimia*, Oxford: Basil Blackwell.

Jackson, P. (1997) Paper presented at the Association of American Geographers Annual Conference, Fort Worth, April.

James, A. (1990) 'The good, the bad and the delicious: the role of confectionery in British society', *Sociological Review* 38: 666–88.

James, A. (1993) 'Eating green(s): discourses on organic food', in K. Milton (ed.) *Environmentalism: the review from anthropology,* London: Routledge.

Lévi-Strauss, C. (1964/1970) *The Raw and the Cooked,* London: Cape.

Lupton, D. (1996) *Food, the Body and the Self,* London: Sage.

MacClancy, J. (1992) *Consuming Culture*, London: Chapmans.

McKie, L., Wood, R. and Gregory, S. (1993) 'Women defining health: food, diet and body image', *Health Education Research* 8(1): 35–41.

McSween, M. (1993) *Anorexic Bodies*, London: Routledge.

Mennell, S. (1985) *All Manners of Food: eating and taste in England and France from the Middle Ages to the present,* Oxford: Blackwell.

Mennell, S., Murcott, A. and van Otterloo, A. (1992) *The Sociology of Food: eating, diet and culture*, London: Sage.

Millman, M. (1980) *Such a Pretty Face: being fat in America*, New York: Norton.

GILL VALENTINE

Morgan, D. (1996) *Family Connections*, Basingstoke: Macmillan.

Murcott, A. (1982) 'On the social significance of the "cooked dinner" in South Wales', *Social Science Information* 21: 677–95.

Murcott, A. (1983a) 'Cooking and the cooked : a note on the domestic preparation of meals', in Anne Murcott (ed.) *The Sociology of Food and Eating*, Aldershot: Gower.

Murcott, A. (1983b) '"It's a pleasure to cook for him": food, mealtimes and gender in some South Wales households', in Garmarnikow *et al.* (eds) *The Public and the Private*, London: Heinemann.

Murcott, A. (1993) 'Talking of good food: an empirical study of women's conceptualisations', *Food and Foodways* 5(3): 305–18.

Payne, M. and Payne, B. (1993) 'Eating Out in the UK: Market Structure, Consumer Attitudes and Prospects for the 1990s', Economist Intelligence Unit Special Report No 2169, London: Economist Intelligence Unit Ltd.

Reynolds, M. (ed.) (1990) *Erotica: an anthology of women's writing*, London: Pandora.

Saltonstall, R. (1993) 'Healthy bodies, social bodies: men's and women's concepts and practices of health in everyday life', *Social Science and Medicine* 36(1): 7–14.

Schwartz, H. (1986) *Never Satisfied: a cultural history of diets, fantasies and fat*, London: Collier Macmillan.

Valentine, G. (1998a) 'Food and the production of the civilised street', in Fyfe, N. (ed.) *Images of the Street*, London: Routledge.

Valentine, G. (1998b) 'Doing porridge: food and social relations in a male prison', *Journal of Material Culture*, 3(2): 131–52.

Valentine, G. (1999) 'A corporeal geography of consumption', *Environment and Planning D: Society and Space* 17, in press.

Vardy, W. (1996) 'The briar around the strawberry patch: toys, women and food', *Women's Studies International Forum* 19: 267–76.

Visser, M. (1993) *The Rituals of Dinner: the origins, evolution, eccentricities and meanings of table manners*, London: Grove Weidenfeld.

White, P. and Gillett, J. (1994) 'Reading the muscular body: a critical decoding of advertisements in *Flex* magazine', *Sociology of Sport Journal* 11: 18–39.

Wood, R. (1995) *The Sociology of the Meal*, Edinburgh: Edinburgh University Press.

Zukin, S. (1995) *Cultures of Cities*, Oxford: Blackwell Press.

13

WHERE YOU WANT TO GO TODAY (LIKE IT OR NOT)

Leisure practices in cyberspace

Caroline Bassett and Chris Wilbert

Real-time environmental control will take over from development of the real space of the territory

(Virilio 1997: 25)

Studying practices displaces some of the ways geography has posited an 'authentic' popular experience against its commodified forms

(Crang 1997: 359)

Introduction

The science fiction writer William Gibson famously coined the term 'cyberspace' in the 1980s, defining it as 'a graphic representation of data abstracted from every computer in the human system', a 'consensual hallucination experienced daily by billions of legitimate operators' (Gibson 1986: 67). A decade later, in the late 1990s, cyberspace's appeal as a fictional construct remains. But cyberspace also exists. It describes that intangible and insubstantial space inherent within the very material constellation of cables, computers, software, and computer operators – users – which makes up the Internet. This is the 'network of networks'; an increasingly fine digital communications mesh which reaches around the globe, connecting some 50–70 million people (Cairncross 1997).[1] The Internet itself, and the associated technologies which stand behind its development, emerged out of the Cold War (Rheingold 1994). Later, the Net was opened up to the academy. Today, it is used as much for commercial, as for any other activity. At nearly all the stages in its history however, the Net has been used for leisure – albeit in different ways.

There are many leisure uses to which digital technologies are put. They include commercial computer gaming (on standalone machines, on playstations, in arcades, or – increasingly – across networks).[2] However, there are also a slew of less organised activities which Net users in particular involve themselves in; activities which are not, as Crouch puts it 'formally labelled for leisure' (Crouch 1997: 189). These include developing home pages, building special interest sites, surfing the Net, participating in electronic communities – for instance MUDs (Multi User Domains) and their chat room descendants, e-mailing friends (and enemies), authoring, reading, writing. It is the quality of these informal leisure uses of technology, many of which began in the early days of the Net, which we investigate here.

The 'impact' of information technologies

We start by sketching out two of the key debates around new media technologies. First is the range of, often technological determinist, arguments around the 'effects' (variously understood as actual or potential, 'good' or 'bad') of information technologies on society. For some, the cyber-optimists, information technologies are understood as benevolent; a forceful force for the good.[3] Cyber-optimists argue that new kinds of activity are made possible by the creation of virtual spaces within information networks, where bodies (and genders) no longer matter, and where geographical distance is collapsed. Digital technology is producing new ways of 'playing together' at a distance. It might thereby help repair the possibilities for a 'community' life – through the development of new public spaces which facilitate new forms of collective and creative activity (see for instance Rheingold 1994). Against such optimists stand the cyber-pessimists – those who see in information technology only the possibility of the destruction of the remaining public spaces of the real world. Information technology, it is argued, replaces the 'real' with the privatised and controlled 'non-spaces' of the virtual. In these spheres mobility – freedom of movement – is exchanged for motility, and touch replaced by a pervasive but prostheticised contact. As Arthur Kroker somewhat colourfully put it:

> Cyber-interactivity is seen as the opposite of social relationships; the human presence is reduced to a twitching finger, spastic body and oversaturated information pump that makes choices within strictly programmed limits.
>
> (Kroker 1994: 23)

Arguments between the cyber-optimists and pessimists are predicated on an assumption that (for good or ill) 'information technology' does have the power to change society – or to create new spaces beyond it. The second key debate here then, involves new protagonists who question this assumption – and the technological determinism that underpins it. At issue is the relative autonomy (or otherwise) of virtual spaces. Some argue the virtual is separated from Real Life – a 'world apart' in which radically new forms of identity can be performed and constructed (Stone 1995). Disputing this contention are those who argue that cyberspace cannot be seen as a 'parallel universe' with its own terms and conditions. Rather, it has to be understood as entirely bound into the terms of the old (see for instance McBeath and Web 1997: 250).

Leisure: the rhythms of work or play?

Leisure itself, of course, is a complex and historically changing concept. Here we draw attention to two facets of leisure. First, it should be stressed that leisure practices are spatial practices. Leisure and recreational practices are one of the main ways in which people make 'maps of meaning' of their everyday worlds (Squire 1994: 12). Second, we stress that leisure cannot simply be seen as free, or even relatively free, time. Leisure is not autonomous from work. Rather the two categories are flexible, mobile and interwoven (Rojek 1995: 171); there are continuing tensions between a sense of leisure as escape, or freedom, from work and an opposing sense in which leisure is seen to reflect and underpin the rhythms of work.[4] Relating this conception of leisure to the question of new machines, we suggest that the pleasures of new leisure geographies are not an escape from work, but are *connected* to changing work practices. It might be argued that we are witnessing new forms, new legitimations and contestations of leisure, recreation and pleasure, partly as a result – and as a response to – the computerisation of work. In fact, as Ross has noted, so far computer technologies have transformed working practices more dramatically than they have re-tooled leisure activities (Ross 1998: 19).

Here, we want to be cautious. Despite the claims of some of those who argue for an 'information revolution' the transformation of work through computerisation is uniform neither in its 'benefits' nor its 'effects'. First, this is evidenced in the gap between North and South – the high tech industries tend to exploit the poverty, cheap labour and lack of environmental regulation of the South, siting manufacturing plants there, largely for the benefit of the North (Golding 1998). For many in the South to work in the information industry means to work on a production line assembling components.[5] Second, in the North itself, the shift to work on-screen, has not tended to flatten class and/or gender distinctions. As Stanworth points out, the flipside of the rise of highly skilled 'symbolic analysers' is the parallel rise of routinised 'information workers' engaged in data entry, tele-servicing or 'dumbed down' computerised checkout work (Stanworth 1998: 53).

Let us now bring together what we have outlined so far. In what follows, we are considering the emergence of some leisure spaces and practices on the Internet in the context of debates around the impacts of technology itself. However, we are also using a notion of leisure as more than free-time to cut across these debates. Using this approach we reject the notion of cyberspaces as autonomous pleasure domes. On the contrary we argue that the dichotomy sometimes imposed between entirely 'playful' worlds emerging in virtual places, and the 'grounded' serious business of everyday life, is a false one. The boundary between these spaces – and the purposes to which they are put – is more complex than that. At the same time, we work with a sense that boundaries between leisure and work, are already pervious – in the 'real world' as much as in cyberspace, and in both cases in 'practical' and 'ideological' terms. Consideration of leisured use of the Net we suggest, underlines the complex relations between 'real' and the 'virtual' spaces; relations which certainly can't be mapped onto a parallel topology of the 'commodified' and the 'authentic' space.

Our starting points are these. First, we focus on the practice of users. That is, the Net is considered from the perspective of practice, rather than in terms of representation. The practices of users, we argue, do not separate, but link the two spaces, the virtual and the real. Our movements within the virtual grounds of new media environments increasingly thread in and out of that portion of our everyday lives, in which we live in 'terrestrial' space. As Sherry Turkle (1996) has expressed it, we *cycle* between virtual and real 'windows'.

Second, we stress another connection with the 'real' world by considering the question of who 'owns' a particular virtual space. We show that both the developers of on-line commercial sites, and the developers of the underlying technologies of the Net, are imbricated within economic, cultural and social orders.

Third, we suggest that the geographies of virtual spaces, such as the Internet, emerge as a play between the practices of (resisting or assenting) users and the constraining architectures imposed by those who own the technological terrain in question. Currently the terms of this play are shifting as commercialisation brings tighter control.

We develop these arguments below via a brief look at two very different moments in the history of the Net, considering first the 'authorised version' of the Net's early history, as told by Howard Rheingold, and other techno-cultural theorists. We then look at a contemporary Net site called GeoCities.

'Free' Cyberspace: The Bridge

> The bridge – aslant through all the intricacy of its secondary construction. The integrity of its span was as rigorous as the modern program itself, yet around this had grown another reality, intent on its own agenda. This had occurred piecemeal, to no set plan, employing every imaginable technique and material. The result was something amorphous, startlingly organic.
>
> (Gibson 1994: 58)

Designed by the US military to withstand a nuclear war, the Internet began as a decentralised communications system based on a principle of redundancy. Scientists using the Net for research began to pass personal messages on e-mail in 1969, while Multi-User Domains were established in 1979 by students wishing to play fantasy games on computer networks, and at the same time UseNet sites piggybacked their networks onto the Internet (Rheingold 1994; Zakon 1998). The activities of these new users exploded the Net, transforming its sterile (if rhizomatic) military geometries via an unauthorised and seemingly unstoppable accretion of additions; a labyrinth of *ad hoc* sites and hyper-links, legal and illegal, inter-penetrating the official structures. Like Gibson's bridge, the Internet was settled, colonised, inhabited, navigated, surfed, and linked, by the practices of its users.[6]

This flurry of activity produced the first flourish of cyber-optimism – these early leisure spaces were said to be 'out of control' in so far as they twisted free, not only of their military origins, but from the constraints of society. The early Net was celebrated as a resilient site for the free exchange of information;[7] for the 'queering' of gender norms (Plant 1996); and for the obliteration of distinctions based on race or class.

Whilst many (correctly) dispute this over-optimistic interpretation of events (see below), it does seem justifiable to argue that users were highly active in this phase of the development of Net geographies. Early Net users were 'walking' a world into being, producing through their textual practices, new geographical spaces – albeit imaginary ones. Where these spaces persisted over time, they often became defined as virtual communities. Indeed for the early Net users the sense of creating a space – particularly a shared space – was apparently exhilarating in and of itself.

Michel de Certeau, the French theorist of Everyday Life, writes of the power of spatial tactics, practised by myriad individuals, to subvert the strategic, the dominant. His examples include the poacher, the bricolager, the walker in the city, all of whom live within but write against – *poach* – the structures they inhabit (de Certeau 1984: 91–110). For de Certeau, walking and writing (each of which can inform the other) are spatialising, resistant, practices, set against the rationalising tendencies of technology. Following his work we might understand early Net users to have been creating subterranean geographies – appropriating pleasure geographies, perhaps, within the grounds of 'military' machines.

To follow de Certeau, however, is to be clear that these would always be temporary geographies; tactical resistance (for him the only possible resistance), always has its limits. It cannot overturn the strategic discourse within which it finds itself. It involves temporary possession; it is a seizing of the time. The *ad hoc* settlement of the Net might be understood as an example of this kind of temporary resistance. In contrast the later history of the Net could be read as a re-assertion of the dominant culture over the new grounds rather unexpectedly opened up by users.

'Owned' cyberspace: the city

> No matter how spontaneous their emergence, self-organising systems are back in organisational mode as soon as they have organised themselves.
>
> (Plant 1997: 49)

The early Net was popularly supposed to be resistant to commercialisation (at least of the 'wrong' kind), perceived as inimical to its (self-proclaimed) 'free-wheeling' anarchic spirit. But it was also vulnerable. As Rheingold put it:

> The Net is still out of control in many fundamental ways, but it may not stay that way for long . . . it is still possible to make sure that this new sphere of vital human discourse remains open to the citizens before the political and economic big boys seize it, censor it, meter it and sell it back to us.
>
> (Rheingold 1994: 5)

Rheingold's fears were justified. In the early 1990s, the Net was transformed, most visibly as a result of the introduction of the WorldWideWeb, and the introduction of Web Browsers such as NetScape. This facilitated an explosion in home use, the effective collapse of the super highway into the Net, and its rapid commercialisation (Ross 1998: 8). In particular, the notion of virtual communities, once the province of obscure recesses of the Net (e.g. MUDs, MOOs, and bulletin boards) became a buzz word for Net developers of all kinds (*Business Week* 1997). Below we turn to GeoCities, an example of a commercial Web operation, and one of the most successful of the later communities.

GeoCities

GeoCities is a place in cyberspace, a commercial site which declares itself a 'community' and invites members to join. So far, over 2 million people have – leading to the boast that this is the largest community in the world. Members are offered a free 'home', receiving a free e-mail account and 11 megabytes of space to host their own page. A further twist is that users are invited to consider themselves as residents of a particular shared space. Users 'place themselves' within 'neighbourhoods' defined by their general interests. Thus, the environs of

186

GeoCities are promoted as a place for leisure – a place to hang out in, to make friends in, a place to 'architect', dream and 'live' in – even if all these activities are marginal to the very tangible attractions of free e-mail.

What users do might appear easy to typologise on the face of it. Once they have settled, homesteaders are declaring themselves *resident* in virtual space, available for *communication* through cyberspace, *competent* in the basic technological literacies, and *creatively* involved in symbolic production. More problematic however is trying to separate any of these activities out. Is writing a home page 'technical' or 'creative', for instance? And is the pleasure of being virtual about a technological competency, or a pleasurable play with identity itself. Users themselves often elide these activities and the fusion also emerges in GeoCities' own rhetoric which habitually conflates the 'technical', the 'creative' and 'communicative'.[8] Another user activity, also often difficult to distinguish from the practices outlined above, is what could be described as digital entrepreneurialism. Users are invited to turn their techno-play into techno-work – to make it productive – or perhaps to regard this work as leisure – in a way which might be said to blur some divisions between the practice of 'work' and that of 'play'.

If it is difficult to separate these activities this is surely because they actually involve substantially similar processes. In these spaces, formally disparate 'practices' (of 'self', of 'communication', of 'selling') could all be understood as achieved through an interaction with the machinic geography of the GeoCities.

Community development

Let us now briefly consider how GeoCities could be understood in terms, not of the possible resistances of its users, but in terms of the groundsmen; those who prepare the pitch.

GeoCities is typical of a new wave of commercial Web development which stresses the production of 'community' – understood to have emerged organically on the old Net but now to be re-tooled in the interests of commerce. Enrolling 'community' in the commercial exploitation of the Net, it is argued, delivers larger groups to advertisers, because people tend to hang around community sites longer (*Business Week* 1997). More importantly, these groups may be self-segmented by particular interest, may be more stable than general surfers, and may provide fairly detailed information about themselves – for all of these reasons they can be targeted in a sophisticated way, and are therefore highly attractive to advertisers (Hegel & Armstrong 1997).

Much has been written on what virtual community is but here, rather than unpacking the notion of community – particularly in its commodified versus (mythical) authentic form – we want to examine 'the place of place' in the development of GeoCities in terms of leisure geographies.

We argue that the production and development of imaginary geographies is an essential part of that push to develop and exploit the connotative aspects of community, in order to produce a place which attracts and holds users, and which controls them. GeoCities, we suggest, functions as a community or can be exploited as a community, through the careful development of a sense of place by its owners; through, that is, giving it a geography which is convincing and which users choose to accept.

Mapping GeoCities

In one sense, the space of GeoCities – used in 'real' ways, for real interactions – is imaginary. A series of addresses and data on disparate servers, it is based on a shared (consensual) illusion of geography. For instance, there is no spatial correspondence between the mappings visible on-line and the order and location of the data on the server.[9] Further, if there is no geography 'of the machine', neither does this geography relate to the terrestrial world. The place-names (Cape Canaveral, Silicon Valley, Athens, etc.) demarcating GeoCity neighbourhoods, do not stand in relation to each other according to the order of a real geography, but as cultural markers for aligned interest groups. Some of the 40 communities indeed skip even this residual reference to a real geography – the eponymous *family* community, with its two suburbs *heartland* and *religion*, is an example of this.

GeoCities is in one sense 'only' a meshwork of sites, a sere architecture of words, images, and locations, inhabited by users – largely invisible to each other. Clearly it is partly these users, adding their own productions, soliciting audiences, involved in their own mail, who bring to the GeoCity pages a kind of life. The consensual illusion of place, of being 'in' a particular place with a location in space and time, may still emerge – in part – 'organically' through the practice of users. However, we'd also suggest that in GeoCities the precipitation of a sense of place out of the airy abstractions of cyberspace, is also a process fostered, developed and managed. In pursuit of 'virtual community' GeoCities is substantially – and methodically – *thickened* by its owners.

First, variety is introduced. The town-grid is made the more fine, through the formal sub-divisions of the 40 neighbourhoods, or 'themed communities', each with their suburbs, co-opted leaders, their newsletters and their specific 'feel'; these are different locations within an over-arching environment. Other tropes, like 'Landmark Sites ' also provide distinctive places to head for, and function to 'landscape' an otherwise fairly blank space.

Second, the organisation of real-time events – live chat and timetabled chat sessions – points time. The existence of a chat stream means that the GeoCities as a whole, largely composed of time-lapsed pages, straggling backwards into the no-time of archival cyberspace (the undated, un-upgraded page) is thus compressed into a 'now' – the now of the individual user at the time of her using.

Third, GeoCities is buttressed by walls of words. These are performative de-clarations of an environment; they include announcements on child safety, on bigoted or obscene writings, on 'trust' (GeoCities is a member of *Trust e.*), the GeoCity 'Privacy Statement', FAQs, and other codes of conduct. These walls result in the production of a wholesome space 'fit' for advertisers – and a particular atmosphere, a defining ambience, for users.

Fourth, GeoCities emerges as a site through increasingly sophisticated branding; GeoCities stamps a floating watermark on user pages – wherever a user nav-igates, it reforms itself at the bottom right hand corner of the viewing screen. The owned environment thus wraps itself around the browsing user, and inter-penetrates the 'work' of the resident.

Fifth, programmes like GeoTickets act as incentives to encourage users to display advertising banners from other users, from GeoCities itself, and from commercial sponsors, in their home pages. The general cross-pollination of GeoCities pages and services achieves two things: encourages traffic flow, and circumscribes it within the GeoCities orbit. GeoCities might be regarded as a form of 'gated community' – albeit one without walls, whose boundaries are delineated by the edges of a flexible environment spread out on the 'inside'. Crucially, the intention is to persuade users to stay in, not to keep strangers out. Users are to be encouraged to 'homestead' – which helps produce the en-vironment, and to surf in circles – within GeoCities. The 'gates' themselves exist in the form of disincentives to leave and incentives to stay.

Territory and environment?

This then, is a map of a particular cyberspace today, a joint production emerging through the practices of inhabitants, practices configured through the solicita-tions of a carefully produced environment. To unpack the relations of the forces involved in this production, it is worth drawing out an emerging distinction between this account of GeoCities and our account of older cyber-geographies. This distinction is the difference emerging between a conception of the Net as territory – as a terrain performatively shaped – and re-shaped – through the practices of mobile users, working against the resistances of the existing struc-tures (which is to follow de Certeau), and a conception of virtual space as about environmental control (the soft cell, pre-produced, and tuned and re-tuneable to the users' needs, at their request). As Paul Virilio put it: 'Real-time environ-mental control will take over from development of the real space of the territory' (Virilio 1997: 25).

To think GeoCities in terms of an editable environment rather than a terrain, has its implications for how users' activity in these spaces is understood, shifting the emphasis from activity to what we might call active passivity, and from mobility to stasis. Virilio's line of thought would suggest that the pleasures of this cyberspace are about a kind of wished for immobility. Increasingly, argues Virilio, technological pleasures are based around a model of dis-ability and constraint, rather than around that model of increased mobility (motor ability) brought about through the 'escape velocity' (Dery 1996; Virilio 1997) which the immaterial grounds of cyberspace seemed at first to provide.

Interaction versus action?

The notion of a movement from territory to environment might be further developed by contrasting the notion of *user action* with *user inter-action*. For Virilio, action is understood as movement within a territory; within the customisable environments of virtual space on the other hand, action – meaningful action – is precisely what has been taken away, only interaction can occur. Here, we might also invoke Stallabrass' alarming description of computer games as an illusion 'not only of scene, but of action'. (Stallabrass 1996: 109). Not only are the contents of interactive games pre-configured, he argues, the on-going responses of users (at least at the time of using) are also regulated by means of the furious activity the interactive game demands. The repeated and frantic actions of the user therefore do not take her anywhere new. On the contrary, these inter-actions pin her down – she must respond as required at all points to move the game along. Stallabrass is writing of 'enclosed' fictional spaces, but a shift from terrain to environment, on the Net of course, might be (and is) understood precisely as a shift in this direction.

Conclusions: flexible formalisation?

The authentic commodity trap

In conclusion there are three points to make. First, we are talking about a process of commodification, but a simple 'before' and 'after' comparison between late and early years of the Net would be invidious. Mike Crang (1997: 359), has pointed to the dangers of what he calls the 'authentic commodity trap' – the ways in which geography has posited 'popular experience against its commod-ified forms' and in doing so, misread both.

We suggest there is a real danger of romanticising 'old' cyberspace by placing it in opposition to its current commodified form. Many accounts of early cyber-space whilst correctly stressing the activity of users, over-stressed the extent of resistance – and have been widely critiqued for it. Claims that the Net was ever beyond the law, or beyond ideology have been exposed as romantic. Theorists and users have under-scored the continuance of gender distinctions despite wide-spread gender experimentation, the cultural dominance of the North, and issues of basic access. Finally, the co-existence of right wing libertarianism alongside the superficial rhetoric of left radicalism, in Net cultures has been noted (Barbrook and Cameron 1998). Facilitated by such libertarianism, commerce has always been evident on the Net. Later Net use – and the sense of the entrepreneurial we have isolated as one of the activities of users in GeoCities – is thus not new at all, in and of itself. It connects with a broader culture of 'enterprise' which has – arguably – always existed within the milieu of information technology.

Clearly aspects of the early Net were less 'radical' than they appeared. And if old cyberspace wasn't utopia, the new isn't entirely dystopic either. When Virilio argues that virtual technology signals the end of any sense in which users may appropriate, or even reappropriate, anything – in the virtual or real world – he is surely over-stating the case.

Earthed bodies

In this paper, avoiding these traps, thinking through the dynamics of developing leisure spaces beyond the terms of a simple authentic/commodity dichotomy, has involved a stress on leisure practices. In this again, we have followed Crang, in part because we find his distinction between representations and 'practices that create representations' germane to virtual spaces (Crang 1997). Considering virtual geographies as representations makes it perilously easy to think of them as existing fairly and squarely on the far side of the screen. Once the practices of users are considered as central however, then users' existences as embodied, remembering, individuals, located on both sides of the screen, moving between virtual and other activities, forcibly come to the fore. The connectedness of the virtual and real then becomes impossible to ignore.

Users cycle rapidly between the virtual and the real over time. Users negotiate a simultaneous sense of self divided across a virtual and 'real' body. More basically they access the Net as located, classed, raced, gendered individuals, and carry that sense into their virtual experiences – where other users, similarly weighed down (earthed) will do the same. The geography of a particular virtual space (owned or not) has its own particular qualities, but a part of these inhere in the fact that every virtual world is shot through with, infected with, the real world at every point in its use. Cyberspace is thus punctuated, saturated, with the values, the ideologies – and the economic constraints – of real life. The commodification of cyberspace, the development of leisure zones within it should not be understood as the penetration of the virtual by the real.

On the contrary, it was only possible, because the Net was always already a part of the real world.

Flexible formalisation

There is then a continuum between old and new cyberspace. However, we have also outlined a series of differences between owned and (and here we amend our own characterisation) *squatted* cyberspaces. Clearly increasing control is being exercised over virtual spaces initially characterised by a general lack of control and – sometimes – by high levels of spontaneity.

We set out to consider informal leisure practices on the Net – e-mail, home-pages, homesteading – rather than the formal arenas of games. However, we would suggest that any distinction between formal and informal spaces on the Net is tending to break down. We would like to suggest that GeoCities, ostensibly an informal leisure space, a space for users to do what they please, and to carry out many different forms of activity, is actually a tightly organised zone, in which activities are highly directed. *The illusion of informality, rather than informality itself, informs this leisure geography.*

It is ironic but not unexpected, that the plastic, abundant qualities of virtual space, which provided the possibilities for creative play, are precisely the qualities which are now being adopted to produce this highly flexible, and by virtue of that flexibility, extremely tightly controlled environment. This is an environment which is not a pre-plotted territory, but which, on the contrary, re-shapes itself around the user, according to their tastes, an environment for aligned and efficient consumption; a consumption in which the user is sold to, and sold. And to which, we should not forget, they consent.

Notes

1 The higher estimate is for e-mail access only.
2 The computer games industry is worth some $14 billion globally per annum (*Screen Digest* August 1998) and is increasingly challenging film and TV as leisure pastimes. There is also increasing cross-over between computer games and more established media.
3 As an early edition of *Wired* magazine put it computer technology was bringing about a 'peaceful, inevitable revolution'.
4 One element of these tensions we do not cover here may indeed be around the desire or antipathy towards speed. As Ross (1998: 20) argues, for every one person who wants their computer to go faster there may well be 50 who want theirs to go slower. This is because 'the speed controls of technology are routinely used to regulate workers' (Ross 1998: 23). On the other hand, in leisure activities it may well be that it is this speed that is welcomed.
5 Cairncross (1997) also observes the move towards outsourcing routine software production to the South.

6   See Ziauddin Sardar's 'Cyberspace as the darker side of the West' (1996) on the negative implications of colonisation/settlement analogies which are often found in discussions of cyberspaces.
7   See the archives of the Electronic Frontier Foundation, and the writings of John Perry Barlow in particular, on this early confidence and celebration.
8   'We allow 11Mb of disk space to allow you to fully express yourself, and support an incredible number of different file types to encourage your creativity' (From GeoCities information).
9   It is possible to 'view' GeoCities neighbourhoods in what is termed 'street view'.

References

Barbrook, Richard and Andy Cameron, 1998, 'The Californian Ideology', http://www.hrc.wmin.ac.uk

Cairncross, Frances, 1997, *The Death of Distance: How the Communications Revolution Will Change Our Lives*, London, Orion Books.

Crang, Mike, 1997, 'Picturing practices: research through the tourist gaze', *Progress in Human Geography* 21(3): 359–73.

Crouch, David, 1997, 'Others in the rural: leisure practices and geographical knowledge', in Paul Milburne (ed.) *Revealing Rural Others*, London, Cassell.

de Certeau, Michel, 1984, *The Practice of Everyday Life*, London, University of California Press.

Dery, Mark, 1996, *Escape Velocity: Cyberculture at the End of the Twentieth Century*, London, Hodder & Stoughton.

Gibson, William, 1986, *Neuromancer*, London, Grafton.

—— , 1994, *Virtual Light*, London, Penguin.

Golding, Peter, 1998, 'Global village or cultural pillage?', in *Capitalism and the Information Age*, New York, Monthly Review Press.

Hegel, Hohn and Arthur G. Armstrong 1997 *Net Gain: Expanding Markets Through Virtual Communities*, Boston, Harvard Business School Press.

Kroker, Arthur and Michael A. Weinstein, 1994, *Data Trash*, Montreal, New World Perspectives.

McBeath, Graham B. and Stephen Webb, 1997, 'Cities, subjectivity and cyberspace', in S. Westwood and J. Williams (eds), *Imagining Cities: Scripts, Signs, Memory*, London, Routledge.

Plant, Sadie, 1996, 'On the matrix, cyberfeminist simulations', in Rob Shields (ed.) *Cultures of Internet*, London, Sage.

—— , 1997, *Zeros and Ones: Digital Women and the New Technoculture*, London, Fourth Estate.

Rheingold, Howard, 1994, *The Virtual Community: Finding Connection in a Computerised World*, London, Secker & Warburg.

Rojek, Chris, 1995, *Postmodern Leisure*, London, Sage.

Ross, Andrew, 1998, *Real Love: In Pursuit of Cultural Justice*, London, Routledge.

Sardar, Ziauddin, 1996, 'alt.civilisations.faq: cyberspace as the darker side of the west', in Z. Sardar and J.R. Ravetz (eds), *Culture and Politics of the Information Superhighway*, London, Pluto Press.

Squire, Shelagh J., 1994, 'Accounting for cultural meanings: the interface between geography and tourism studies re-examined', *Progress in Human Geography* 18(1): 1–16.

Stallabrass, Julian, 1996, *Gargantua: Manufactured Mass Culture*, London, Verso.

Stanworth, Celia, 1998, 'Telework and the information age', *New Technology and Employment* 13(1): 51–62.

Stone, Allucquere Roseanne, 1995, *The War of Desire and Technology at the Close of the Mechanical Age*, Cambridge, Mass., MIT Press.

Turkle, Sherry, 1996, *Life on the Screen*, London, Weidenfield & Nicholson.

Virilio, Paul, 1997, *Open Sky*, London, Verso.

Zakon, Robert, 1998, 'Hobbes' Internet Time Line', http://www.isoc.org/zakon/internet/history/HIT.html

14

THAT SINKING FEELING

Elitism, working leisure and yachting

Eric Laurier

Finding a footing

How shall we know what we are feeling? An everyday question, yet a perplexing question nevertheless. Knowledge and feeling remain held in opposition and knowledge is all too often in the dominant register (Averill 1974). *How shall we feel what we are knowing?* This second question is seldom on our lips. To be candid, I doubt I will do any justice to this second question in the chapter that follows. Instead I shall be reaching toward knowledge of the emotions, but if you the reader can hold onto the second question then you will have a way of disagreeing with my feelings from the very outset of this journey around the coasts of work, leisure and elitism. Leisure has been defined in many ways by many different researchers and I am going to try and define it as not one but several ways of feeling. Then this suggests a third question that I should try and answer: *how can you (or I) feel that way?*

In what follows I will try and tack between the rocks of 'personal' experiences and the shoals of a set of social theories. This tacking between two different ways of knowing and feeling the world will be reliant on a set of navigational aids. To find your way in what follows will involve relying on marking from the outset the critical importance of the acquisition of inequitably distributed competencies and dispositions in leisure practices (Lamont 1992), the learning of practical knowledges (Bourdieu 1977), and the embodiment of resources ('finding your feet', 'knowing the ropes'). Knowing the boundaries by assisting in the production and reproduction of their lines of division and bounding the known by assisting in the learning and teaching of how to be at leisure. This ethnography provides a kind of counterpart to Willis's (1977) classic *Learning to Labour* which examined the enculturation of working class boys, in this case it is about the upper classes *learning to leisure*.

What does it mean to critically reflect upon how we attempt to amuse ourselves? It might mean that we spoil our own sense of fun entirely, which would in turn remove the licence to claim that we are able to feel what we claim to know. We would only know *about* leisure and no longer just *know it,* or of course in the words of one particular leisure brand, *just do it*. All of which shifts us gradually toward some very geographical questions about the making of boundaries and their materialisation in yachting as a leisure practice.

Moorings

'May I come aboard?'   'Yes, come aboard.'

> *And so I made my nervous leap from the harbourside onto the restored 1920s yacht. The elderly couple had been waiting for me for about an hour or so because of my overly-ambitious estimate of how long it would take to drive to the harbour. As soon as my feet hit the deck and my bag thumped down beside me, Hugh (the husband) set the motor running and the boat pulled away. He was irritated already because we had almost missed the tide, Dee, his wife, explained ushering me into the cockpit. Not exactly a promising entrance strategy, I thought and then noted with a shiver that I probably hadn't brought enough warm clothing.[1]*
>
> *I could not pretend to be a total outsider to sailing since I had had a GRP dinghy of my own for several years as a teenager and my father had been dragging me out on small boats since I was a toddler.[2] I had even spent a week on a sail training course learning the basic principles of sea, wind and sail. For all of that experience I was still gripped by a great deal of anxiety as I stowed my kit in the main cabin and felt the first offshore waves tipping up my feet.*

Although this chapter will follow the narrative of just one boat and its occupants during a short trip along the coast, it will shift out of its local account to reflect upon wider relations between leisure, knowledge and elitism.[3] Briefly, it will drift away from leisure as simply a conservatively defined public good – 'playtime of the masses' where it creates social cohesion, self-realisation, bonds of community, responsible citizens, healthy and committed individuals (Kraus 1987), in other words leisure as 'an all round good thing'. This drift begins by considering just who is able to 'play' at sailing since only 1.2 per cent of all households in the UK own a sailboat of any kind let alone a large yacht (economically yachts are the largest part of the boating sector but numerically dinghies dominate, see Mintel 1997). Many more people aspire to own or sail a yacht, since it occupies a particularly powerful symbolic position in both the signification of social status and the imagination of leisure. Rather than reducing my analysis of sailing to spending power I will try and critically interpret the workings of this particular leisure practice in creating orders of people and things through boundary and emotional work. The investment of money in yachts does not explain, apart from very simplistically, the investment of other forms of capital and the investment of meaning, and intense human feelings in and around yachting. Articulating the passages of these investments are smaller and larger assemblages of machines, knowledges, actors and places which serve to define the power relations of its field (definitions which, like wooden boats, are never in the end watertight). Whether these assemblages are small or large depends on the practices of those involved, whether they 'raft up or anchor off'.

One first monetary grasping of the supportive structure of yachting we can label as the 'pleasure boat sector' (comparable to the classic car sector); though this does not do full justice to the articulation of producers, consumers, sites and trajectories involved, it is the sufficient hook for a series of business interests (Mintel 1997). Reading the texts of market reports on yachting can provide an unsettling reminder of their particular grasp of 'leisure'. 'Sailing is the preserve of ABs who are three times more likely to own a sailboat than the C1s . . . The *good news is* [my emphasis] that ABC1s are expanding rapidly as the middle strata contracts and society polarises along socio-economic lines' (Mintel 1997: 5).

Classic boats are more exclusive since they require greater competence in practical and historical knowledges. Their arrival as a popular category amongst the affluent classes came in part from the recession in the fishing industry during the early twentieth century, which freed up wooden fishing boats, and also from the ageing of purpose-built river pleasure boats, cruising and racing yachts from the turn of the century. Like the car industry, the pleasure boating industry has been driven in part by product renewal, fashion and technological change and it has created a corresponding lineage of material culture. Cruising and racing are two quite separate practices, though they may involve shared actors, sites and equipment. This chapter will concentrate on cruising as the more 'leisurely' of the two practices for a number of reasons, including: the absence of formal competitive rules that are found in racing; its discursive construction as free time and escape; and its reliance on abstract geographical knowledges (charts), textual maritime orientations (cruising guides), social exchange (dialogues between sailors), personal memory and creative acts (doing your own thing).

Rope knowledge

We didn't go very far on the first afternoon. I think Hugh wanted to watch me struggle with some ropes and sails for a while to assess my competence on board and having realised I didn't know much he moored the boat and left the difficult passage through Cuan Sound (a tidal passage on the West Coast of Scotland) for the next day.[4] For the most part he maintained a grumpy presence at the wheel, from time to time issuing short instructions. Dee was

197

delighted to have me there, and I was pleased that she was pleased while worried that we were getting on so well that Hugh would very quickly question my disposition. My reasons for giving up sailing as a teenager were in part due to its male domination and culture of beer and bravura.

Below decks while Hugh was busy tying knots in ropes above, I asked Dee about Hugh's health since part of the reason I was there was because he had suffered a heart attack a few months previously. 'Oh he's not had any more problems. It's just . . . and it's not that he will . . . but if he did have a heart attack then I couldn't sail the boat by myself. And help would not arrive immediately.' And then she began to organise the cooker and the food while I asked a lot of questions about who does what on the boat. I wanted to see how she justified the very traditional roles that seemed to be taken on board yachts. Women down in the galley cooking, or lying on the deck reading a book, or being told to 'grab that rope there'. She put up with my questioning the obvious in a way that I suspected Hugh would not have. With considerable tolerance she corrected me on my grosser mistakes; she was a capable sailor but it was like when they drove together in the car, Hugh wanted to be in the driving seat. As the weekend wore on Dee eventually told me a story about when she had first gone sailing with Hugh as a young woman. She was at that point 'quite a wild young thing' and 'had several men on the go at the same time'. Hugh was not the most charming of them but somehow he had convinced her to come sailing with him for a weekend. Being confined on a boat with him had made her nervous at first (a feeling I could understand) but his competence and ability on board won her over. She realised that she could place her trust in him completely because she never felt in any danger from the sea while they were together. In its unhesitating and well-plotted delivery I knew that Dee had told that story about placing her trust in Hugh many, many times before. It was a key narrative in her fashioning of their marriage and of Hugh's value as a man.[5] And in the context it was told, as Hugh barked orders, steered through rip tides and interpreted sonar readings, I had to position myself in agreement with her map of mattering. Though of course my choice of Hugh at the helm was only for a weekend, Dee had left him at the helm for the last thirty-five years. Or had she? Was she taking control in the places that fitted her disposition?

After dinner I asked about whether Dee and Hugh thought of themselves as part of a leisure class. I had tried to lay some ground for this question by chatting a little bit about my research interests and about what leisure might mean. Nevertheless, Hugh decided to leave the table immediately. Oops. How did I upset him? Dee explained that Hugh didn't like explaining himself. So Dee spoke about the way they lent their boat out to several of their friends who could not afford a boat because they were too young or just did not have the money.

'But they knew how to sail of course?'

'Of course.' Dee added that there were several charitable schemes for taking 'disadvantaged' and 'disabled' people out on sailing holidays. I couldn't help noticing the gap between the 'deserving' poor and the rest that simply could not afford to go out on the water. At a later date another yachting enthusiast explained to me that people took shares in yachts, he had known up to ten people sharing the cost of a yacht and then time-sharing it during the sailing season. Most of them, he added, were members of the local sailing club, rather than the Royal Northern Yacht Club. Ownership was not the most important thing anyway, he continued, anyone that really wanted to could find a place as crew on a yacht. There was a constant demand for young people willing to crew. Meanwhile I was busily noting down – 'young people, not old, and mostly young men who don't want to sail with their fathers anymore. It's a way of acquiring the emotional and tacit competencies for the financial power that will arrive later. The dispositions are already in place.'

What is required to understand how elites perform and reproduce competencies is an attenuated account of their life-worlds and critically an understanding of their socio-economic, cultural and moral boundaries, which are worked through in their ways of acquiring and transmitting their *embodied dispositions and spatial knowledges*. To begin to think about how an elite is able to function without declaring itself as such (Parry forthcoming) leaves questions open about whether the term should be used at all. However, there are boundaries drawn around and through the sailing of classic boats (Lamont 1992) which create a passage for the reproduction of elites which is not in the obvious sites of education, family or work. Or perhaps it is the reproduction of an attitude; elitism? To call it that is to risk collapsing it into another disposition and one which most of the sailors I had met would deny. They prided themselves on their egalitarianism and yachting was a resource for other ways of social being that were not based on their position at work or their home address. It was, they claimed, more about a shared *enthusiasm* than a shared *exclusivity*.

Paradoxically, what I also wish to pick up on in a somewhat Foucauldian sense are the various forms of disciplined labour required to mess around a yacht, the disciplines of the practice that prevent mutinies, sinking and the 'wrong kind of people getting involved'.[6] Though it is a disciplined leisure, I still do not want to let go of its *emotional geography* or *map of mattering*.

These 'mattering maps' are like investment portfolios: there are not only different and changing investments, but different intensities or degrees of investment. There are not only different places marked out (practices, pleasures, meanings, fantasies, desires, relations, etc.) but different purposes which these investments can play. Mattering maps

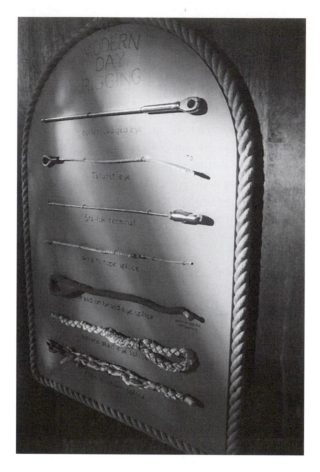

Plate 5 A deconstructed display of rope knowledge

define different forms, quantities and places of energy. They 'tell' people how to use and generate energy, how to navigate their way into and through various moods and passions, and how to live within emotional and ideological histories.

(Grossberg 1992: 82)

Dropping anchor on the second evening I assisted Hugh by peering over the front of the yacht and warning him of rocky outcrops under the water. He had sent me there calling out his credo that 'there must be a system' and his system for finding a mooring in a bay like this one was to use his sonar to calculate an average depth for him, to get me to look for underwater rocks, for him to glance at his charts from time to time and exercise his local knowledge of the island we were coming into. With all these calculations going on, more than usual really because it was low tide and Hugh had chosen a cove with a tricky route into it, he was also keeping one eye on the weather. As I wrote cryptically in my journal: 'it is sooo complex.' This was not the soccer player 'on the ball' that Merleau-Ponty uses as a metaphor to exemplify practical sense.[7] How to play a good of game of soccer could be learnt with the minimal of material resources — a ball, a flattish surface, some goalposts and some other players — which was why it was the democratic if gendered example that Merleau-Ponty used. Sailing a large yacht into a shallow cove required, well to begin with, 'a large yacht', then; charts, sonar, an obedient human assistant, the shipping fore-cast on the radio, weather forecasting skills, a diesel motor for the little manoeuvres, tactile knowledge of how the boat 'handled' and the memory of this island being a 'good place to stop on a night like this.' Hanging over the front of the boat I was filled with the same awe that I had had as a child when my parents got us all into a car and then drove from a suburban house in Scotland to a camp-site in Brittany. How did they do that? How do they find the way? How utterly dependent am I? Where did they learn? It was as great a mystery to me as the nomad's spatial knowledge of the apparently smooth space of the desert is to the anthropologist.

The leap between my skills and knowledges and Hugh's was almost as great as that between the five year old child and his parents taking him on holiday. Except I was an adult now, learning to drive a car to my holiday destination was a practice I had acquired along with the majority of Westerners. Sailing a yacht to take my holiday? I had a sense of how Hugh did what he did and found the way to this cove but I also knew that it was beyond what I could do. The questions I was asking now as an ethnographer were what boundaries had I crossed already to be here and what further ones remained before I could take my leisure as Hugh was taking his (still rather grumpily if I may say so.) In the meantime I retreated to the warmth below decks to peel potatoes for Dee who was making dinner.

Working toward leisure

Do leisure and work stand in opposition to one another? Leisure theory makes it quite clear that they are utterly entangled (Rojek 1985; 1995; Cohen and Taylor 1992; Shields 1992). Release, escape and freedom from work in leisure are unattainable states, yet they remain key signifiers in discourses of leisure. And for elites in particular, as Veblen (1925) argued at the outset of research on leisure, their lavish expenditure and display of idleness marked their distance from necessities of labour for survival in capitalist societies. For Veblen the dreams of escape for all of society were shaped by a leisure class. Something of this remains in the mythology of the yacht, still frequently invoked as the purchase to be made after winning a large sum of money or working up to it, marrying into it or retiring to it.

Perhaps there are moments when a yacht fulfils those dreams and those on board feel free, relax and forget all their worries. Yet to make those moments requires all kinds of work. Drawing on theories of social interaction (Goffman 1967; Gieryn 1983; Hochschild 1983), class (Bourdieu 1984, 1990; Lamont 1992; Skeggs 1997) and embodied performance (Morris 1995) I will make a 'system' of the *workings* involved in cruising on a classic boat into two sets of actions.

Boundary work

Perhaps most important to any argument about elites is that boundaries are always being shaped to include a privileged few and exclude an under-privileged many. Such boundaries pre-exist those that find themselves on one side or the other. For their existence, however, they rely on being reproduced constantly. They are not simply 'symbolic' boundaries in that they are constituted from combinations of socio-economic, cultural and moral measures and they are solid-ified in the material culture. So it is that a classic boat requires considerable capital for its purchase, maintenance, equipping, berthing and sailing. Further, it requires an education in maritime heritage and history, taste, aesthetics and oddly enough a series of moral judgements about integrity, honesty and trust. And when a classic boat is brought into the flow of leisure practices it is expected to assist in their bounding. Boundary work is vital to the creation of conformity, since without 'bounds of reasonable behaviour' then there can be no *deference via embarrassment* (Goffman 1967). As has been suggested by social psychologists, 'it is not hard to see that this account construes embarrassment as an emotion of social control' (Parrott & Harré 1996: 40). It is the 'emotional' character of embarrassment that takes us into the next category. What is worth commenting on here is that being embarrassed because of acting out of context, relies on a shared knowledge of what the context is and where its boundaries lie. Like shame, guilt or chagrin, embarrassment puts an individual's competence on the line and will frequently lead to submission to the limits on behaviour or avoid-ance of the situation where the embarrassment occurred.

Emotional embodying work

Frequently under-estimated in research on leisure, a high degree of emotional management is required by those involved. Leisure is constructed around notions of appropriate emotions, 'having a bad time' is the ultimate failure of precious investment given over to leisure. On the self and for others a considerable amount of effort is thus required around lines of hosting, relaxing, encouraging and enthusing, in other words strategic surface acting is required to maintain the appearance of having a good time together to maintain the definition of the situation. Also, finding out how to feel works in combination with the boundary work since shame and embarrassment, and often fear are powerful cues and 'clues' (Hochschild 1983) for the adjustment of interactions. For most people embarrassment is also displayed by their flesh, in a rush of blood to their face and their becoming flustered. *Avoiding embarrassment is one of the key motivations in normal social interaction* and the encouragement of free acts during leisure activities by increasing a situation's informality multiplies the possibilities of an embarrassing occurrence arising. This is not to say that there are no rules for leisure time. The display of embarrassment in various leisure situations indicates to those involved that there are social conventions and norms framing behaviour and that the embarrassed individual knows that he or she has just found themselves 'faulted' by them (Goffman 1967). On board a boat considerable amounts of practical skills are required, these are more than just getting about, they are also to do with intimacy, 'knowing the ropes' and giving off embodied impressions of enthusiasm.

Times of leisure are thus testing times, where away from the written exams and the other assessments of educational authorities, the work that is done on boundaries, through an emotional body is assessed. To enjoy the trip on a yacht, can the participants fit in financially, judgmentally and aesthetically with their hosts or crew? Can they carry out the many tasks of tying, untying, folding, helming, staying upright and all the while smiling? Can they perform their enjoyment?

> *'Cheer up, there's a good chap' Hugh said to me, his grumpy mood replaced by a chuckling elation as the wind got up, waves started splashing across the decks and his crew took fright. I was hunched over in the cockpit out of the wind looking distinctly unenthusiastic. As I noted later, 'fear was my way of feeling incompetent'. A competent sailor isn't startled by a bit of sea spray and the boat leaning over. With Hugh standing over me I definitely felt his inferior and I just bet he felt damned superior showing his courage off. It wasn't spontaneous courage though like my spontaneous fear, Hugh's courage was learned. It was a socially distributed emotion. Yet the imperative for me I was aware was to at least wrestle my clearly visible fear into an almost camouflaged anxiety.*[8]*
>
> *Much mockery was made of 'fair weather sailors' at the yacht club. They were characterised as people who used their boats for drinking gin on sunny days and only left the marina in a flat calm. They were not 'real' sailors and were dismissed on many occasions as having wasted their money on boats. I surmised that it*

was wasted money in the sense that they would be excluded from the privileges that being a member of the classic boat owners brought. The informal resources: the friends in powerful places, the tips on investments of various kinds, the intro-ductions to exclusive networks and 'movers & shakers'.[9] Or was it just that they failed to experience the thrill of helming a boat through rough seas. Internally, as I took Hugh's hand up on to the deck, I was struggling with a degree of fear knowing that I should try and flatten the very tight fold in my self around lack of control and, well, drowning at sea. 'This is just a light breeze' Hugh added and waved to another boat passing in the opposite direction with its spinnaker up.

I liked this waving-thing that people on boats did to one another. At the same time I was aware it was another one of the parts of the performance, people that did not wave from their boats were 'anti-social'. It reminded me of some-thing that David Crouch had written about caravanners signalling one another in various ways when they were in motion as well. They were making a boundary as they waved and placing each other inside it.

What surprised me about the classic boats when they were at the harbour-side or in a marina was how welcoming they were to visitors. Perhaps this was more a feature of festivals and regattas, Goffman's 'actions spaces' that I had been to, where nearly everyone there would be classic boat enthusiasts. There was something about 'enthusiasts' and 'enthusiasm' that meant more than all care-fully constructed liminal practices I was thinking about. Or was there, since enthusiasm was emotional work? Compliments were paid on the restoration of the boat, the ultra-compact arrangements of life below deck. Interest was always shown toward the leisure object and its accompanying texts (talk, logs, original documents) and displays (photographs, charts of voyages, sketches and paintings). There were few dis-interested visitors. And there were also few who didn't say 'May I come aboard?': the secret password that allows seemingly anyone on to any boat.

Working knowledge or leisure knowledge?

Hutchins' (1996) fieldwork aboard a United States Navy ship showed how cog-nitive skills are unevenly socially distributed amongst the members of the crew. What was of further significance in Hutchins' work was his attention to the place of instruments, charts and equipment as well as knowledge of informal conven-tions aboard naval ships in stabilising hierarchies and ranks at sea. Practical knowledges, according to Hutchins, are acquired by 'kinetic mimetics' as well as by talk, in other words by doing as others did at their level as well as doing what they were told by higher ranks. In its combination of material and dis-cursive orders the acquisition and exercise of practical knowledges on board a 'leisure' sailing boat and the importance of both giving and obeying orders make this form of leisure very appealing to organisations that want to increase the relations of order and obedience in their 'teams' (i.e. during business adventure weekends). Boundaries have a degree of stability in the materiality of the yacht. Each time someone steps aboard the boat they are offered a position which they can then best fit according to their dispositions and competence. If they fit well then they will do well and may even relax as they find themselves comfortable (i.e. unembarrassed and free from anxiety) in that position.[10] While the classic boat's mythology encourages further affective investments from organisational workers that their office as context for passionate relationships is for the most part unlikely to articulate. Order can be embodied in the leisure object.

Much of this discussion remains internal to the space of the yacht, but what of the larger flows of materials and knowledges within which they move? There is a great deal of cross-over between producers and consumers, many of the shipwrights, marina developers, and boat charterers have entered these industries through involvement with sailing as consumers before shifting into the sector as workers. In effect, they learnt ways of consuming before ways of constructing in the maritime realm.[11] In making and selling high unit cost products the close integration between producers and consumers is vital as one unsold boat is a large loss.

With such a high profile collection of consumers, there is considerable work going on by marina developers to commodify the spaces of the coast. These vary from waterfront schemes associated with inner city urban redevelopment, often replacing docks destroyed by the shift to containerisation (Boyer 1994 and Sekla 1996), to attempts to use cruising guides to locate 'green coast' developments. Craobh Haven was a purpose-built marina which involved building timeshare properties, chandlery shops and other facilities, simultaneously a convenient location for yachts cruising the West Coast of Scotland to stop off. *A commodified 'rafting up'.*

Hugh and Dee put great store by their local knowledge of secluded coves like the three different ones that we moored in.[12] *Many were off uninhabited islands and I was able to take the tender (a small rowing boat) and row ashore to explore and get away from the excessive physical and emotional intimacy of the boat, my own unease over dressing and undressing rituals, sleeping arrangements and boat-bizarre toilets; all telling signs of my lack of practical knowledge. And the bruises of course from constantly cracking my head, striking my shins and catching my shoulders on various protruding bits of wood and metal. The way being an observer in this case was something of a misnomer since inhabiting the boat was beyond 'gazing upon it' as a touristic ethnographer. As well as working on my fear of sinking in a storm or hitting unseen underwater objects or being hit by Hugh for incompetence I had to work on my co-ordination in a tight space.*

Hugh and Dee also put great store by their orderly knowledge of the multiple internal divisions, the poetic nooking and crannying of the boat in stowaway boxes, drawers & shelves: 'without these "objects" and a few others of equally high favour, our intimate life would lack a model of intimacy. They are hybrid objects, subject objects.' (Bachelard 1964: 78). Hugh: 'A place for everything and everything in its place. According to a system of course.' My first glance at the inside of the cabin had suggested a kind of chaos but the more I lived in it and watched one pattern of chaos transforming into another after the stowing of morning things and the unpacking of mid-day things the more I realised that it was simply that I was used to things being arranged with gaps between them. Empty space was a luxury which could not be afforded inside a yacht. And it was Dee that was the librarian to this archive of everyday goods. Hers was the work of distilling a house's equipment down to its purest elements and still being able to surprise with a packet of chocolate biscuits from inside a welly boot.

The charts that are used are, in their sparse cartographic details and absence of advertising, relics of an older kind of mapping. Cruising manuals from the 1940s are still used somewhat riskily to guide people around the 1990s coast. And word of mouth from other sailors when 'rafted up' or settled into a harbour-side pub or restaurant provides further guidance on where is 'good to go to' and where is not. These spatial practices and knowledges allow the people on board their boat to move in and out of commodified places. Those variable and shifting articulations are, it seems to me, the kind of privileged freedom that sailing does seem to offer, the power of disconnection from an extensive network that Hincheliffe (forthcoming) identifies in the urge amongst Danish householders to use stoves and fireplaces instead of central heating. Out on the water, cruising in a sail-powered boat, a considerable amount of work is certainly required to keep things moving and yet this work achieves the effects of escape from other kinds of work and larger 'systems' that are beyond the individual. Hugh's 'systems' in the earlier section are *his*, they are not the systems he was submerged in at his workplace before he retired.[13] Even as a manager in the shipbuilding industry, or perhaps more so as a manager in the shipbuilding industry, he felt trapped in 'forces beyond [his] control'.[14] The yacht gives the impression of being a discrete object, a sealed up place independent of larger human assemblies. That is part of its exclusionary appeal.

I got up well before they did so that I hoped no one would hear me peeing over the side of the boat. Even though this was what I had been told to do, I wasn't so sure about it. It was about 7 a.m. I suppose, the water absolutely still — once I had finished rippling the surface that is. A few curlews called out, I could see a white farmhouse on a distant hill, a high grey sky, some whisps of cloud caught on the barbs of a pine forest. Muffled human sounds from inside the cabin and the smell of rain not long finished on the wood. I think I might even have been enjoying myself.

Knowing how to feel about it all

In narrating one episode from a pilot study for a research project on elite leisure practices, I am still working on how I should have felt about it all. The research proposal was a very emotionless document, competently hiding the frustrations involved in writing it and trying not to encourage amusement in the suggestion that studying an elite leisure practice would involve crewing on a yacht. Doing research carries its own special modes of comportment (Cook and Crang 1995; Cook 1995; Parr forthcoming) which are about making boundaries and embodying emotion. Beware the moral boundaries of representing the fun that can be involved in doing research. Even in a research milieu that has been thoroughly reshaped by feminist critiques of heroic masculinity and objective knowledge, stories of suffering or stoicism during fieldwork remain the disciplinary norm.

Studying elites as a 'critical social researcher' sets up a series of socio-economic, moral and cultural dilemmas. Can you afford to do it? What is also on issue is to what extent you are going to be 'genuine' during the time spent with them. Amongst the elites studied by Lamont (1992) identifying phoneys, social climbers and *salauds* was a constant purifying activity to ensure moral order amongst the upper-middle classes — so certain attempts to mimic the dispositions of elites or make tactically attractive impressions may result in your prompt disposal from their company. So the dilemma is a mixture of the interaction dilemmas of impression management and the relatively 'weak' position of the researcher. For me there was still the lingering moral sense that somehow presenting an inauthentic self was an unacceptable practice, even though necessary to gain the confidence of certain dis-similar participants. Added to which, of course, all the manuals on ethnographic work tend to urge a sympathetic treatment of all research subjects (i.e. Kleinman and Copp 1993).

On a cultural level, where are you coming from? Will you be treated as an outsider by the elite because of your own cultural traits, your maps of mattering? In my pilot work I came from the elderly couple's home town, was middle class enough and was enthusiastic enough about yachting. Except I was not really competent enough; yachting in practice frequently involved hosting guests who were neither emotionally nor practically competent and sometimes managing

207

guests who might well end up having a terrible time at sea. For all my fearful moments I was, like the weather, relatively calm during the weekend, showed willing to try and do things I did not have a clue about and my hosts in turn made it all easy because they were old hands at looking after fair weather sailors.

Then there is the final tricky issue that this chapter began with, can feelings be mapped by knowings? Kleinman and Copp's (1993) excellent work on emotions and fieldwork in qualitative research taps into the diverse re-workings of feeling required to engage fully with participants in research projects. They emphasise the multi-layered nature of feeling, the way in which during an interview the researcher can be bored, anxious and curious at the same time, each layer masking the connections of the other layers. Taking their cue from symbolic-interactionism and Hochschild's (1983) work on 'feeling as clue' they stress that emotions cannot be divided off from research and that they are part of the material for analysis. After detailing a great deal of negative feelings encountered during research such as avoidance, despair and anger, they shift on to the good feelings of pride, identification and rapport. They warn of the tendency 'to bask in our feelings of competence, believing that the good relations we have established reflect our skills at developing rapport' (Hochschild 1983: 46). In other words, taking our leisure when things are going well in research rather than asking why things are running so smoothly. Part of the work that leisure requires, as I noted earlier, was in making guests feel comfortable, making the guest, even if he or she is a researcher, feel good and in that sense my enjoyment was part of the experience I had to reflect upon. It was not entirely my skills as a researcher that had led to my being able to bask on the deck of Dee and Hugh's boat during a sunny afternoon. Equally my moments of embarrassment were clues about where there were boundaries being broken. From my experiences of yachting I would like to push the definition of social ordering through emotion on to feelings of fear and courage as a socially acquired response to those feelings. Leisure in these terms provides spaces for the re-shaping of emotions other than pleasure and enjoyment.

In extending definitions of empirical material in research to include the very subjective sphere of emotions, Kleinman and Copp (1993) ignore an important aspect of Hochschild's (1983) ground-breaking exploration on the colonisation of human emotions by commercial ideas of good service. Hochschild's central thesis was that working on our emotions has become a central part of labour in late twentieth-century service-oriented economies. In the search for perfect customer relations, for a kind of 'feelgood factor' so important to profit in sectors such as the leisure industry (Mintel 1997), the repertoire of feelings and their performances of service workers have been delimited in favour of a constant drive towards customer satisfaction and comfort. 'It is from feeling that we learn the self-relevance of what we see, remember, or imagine. Yet it is precisely this precious resource that is put in jeopardy when a company [or research process] inserts a commercial purpose between a feeling and its interpretation.' (Hochschild 1983: 196).

In addition, for the growing number of social researchers who now carry out in-depth interviews, participant observations, focus groups and other forms of affectual research, managing emotions while doing research has become a central part of 'the job'. This means that at the end of an hour's, day's, week's or nine months' work the ethnographer is left working overtime trying to reclaim their inappropriate feelings from their appropriation and suppression in favour of knowledge production. Which is another way of saying that investigating leisure involves knowing what you are feeling and then being able to return to feeling what you are knowing.

To finish on an emotional topic I would like to re-consider my efforts to work out Hugh's grumpy demeanour, since I have commented on my own fears earlier. In interpreting his attitude the premises I began with were: since cruising was Hugh's main source of pleasure then he would be happy and since he was grumpy there must be a reason for it. That reason, I suspected, was my presence on board his boat. So I tried to assess whether I was being intrusive as an incompetent sailor, culturally distant by my class origin, age and related tastes, a social researcher invading his private life, or combinations of these.

My discomfort in his presence was my 'clue' and my required emotional response was to try and establish some kind of rapport with Hugh (they some-times describe it as 'making friends with the enemy') even though I normally would simply have avoided the situation entirely (Kleinman and Copp 1993). After all, Hugh's name was on the list of commodores at the yacht club I remem-bered sneering over as a teenager. So I kept smiling, I kept trying to be bouncy and enthusiastic. Hugh's grumpiness masked my seeing his freedom and perhaps more significantly masked the major emotional work that Dee was doing in the background. My efforts would perhaps have been best directed to attending to her efforts to make me feel at home.[15]

Acknowledgements

Without question my understanding of cruising was thanks to the understand-ings of 'Dee and Hugh', and also several other woody boat people who shall remain nameless — 'Jenny, Will and Robert'. Ian Cook for his own take on being grumpy. And to the Dept. of Geography, University of Wales Lampeter for financing 'Bristol and the Sea' out of which this pilot was made and the Dept. of Geography, University of Bristol for hosting me for a year.

Notes

1 The sections in italics are based on the author's journal kept during the weekend – repeated or emphatic statements from the participants are in quotation marks. It has to be remembered that these quotes were noted down sometimes hours rather than seconds after they were actually spoken – thus they are unreliable and over-authorised 'voices' from 'there'. The paper will proceed in two modes of expression, one more novelistic – in keeping with its basis on the style of the research journal and the second giving a straight-up write-up – based in part on the structure of the abandoned research project. For comparable stylistic approaches on very different subject matters see Anna Wynne's (1988) bi-modal account of MS sufferers and Margery Wolf's (1992) tri-modal writing on Chinese peasant life.

2 GRP (glass re-inforced plastic) dinghies are mass produced and relatively cheap with little maintenance required during the winter months unlike older wooden dinghies. Before the advent of GRP boats the main earlier attempt to make sailing accessible to the lower middle classes was through kit boats. Boat designers like Uffa Fox had produced plans like those used by dressmakers which involved similar methods of construction – sewing together planks of plywood rather than fabric, then gluing with epoxy resin – which sailors used to build their wooden boats in their garage, garden or kitchen (in my father's case) at relatively low cost. The disadvantage with these wooden kit boats was the extra demand on acquisition of competencies (i.e. woodworking and painting skills) as well as sailing and then the ensuing maintenance requirements every winter.

3 Are relations between leisure and elitism 'wider' than the relations Hugh and Dee were making as they sailed around the North Atlantic, the Caribbean and the Danish archipelago or as Hugh and Dee brought together their family, other families, all kinds of equipment from GPS, to traditional canvas, to canned haggis? And how wide does such a relation beome when their leisure practices were reproduced from coast to coast to coast. My theoretical survey will likely be 'narrower' but will still try and neatly arrange distant things into a little chapter reading machine.

4 A typical practice identified by Purves (1996b): 'More difficult are the skippers who stealthily try to find out if you are going to be any use. A favourite ploy is to give you the helm immediately . . . to see whether you'll push the stick the right way. They also long to find out whether you are going to be a liability. As another frank host said on a windy night: "Look, if you can't cope, just for god's sake find your own bucket, wrap up warm and keep out of the way." Ashore, he was the soul of *politesse* and chivalry.'

5 Hugh's competence was socially reproduced; having learnt how to sail a large yacht almost single-handedly from his father and his father's friends. The boat we were sailing on had originally been sailed new-built by one of the people who had taught him to sail. Without wishing to overstate it, most of those who sail yachts come from yacht-sailing lineages which is where they acquire the 'taste for it' and the competencies to be able to do it. The tastes and the competencies are thus unevenly socially distributed and reproduced predominantly through bonded relationships of kith and kin.

6 Reading through the visitor book that Dee and Hugh kept on board I was given character sketches of a couple of the best and worst of their guests.

7 Merleau-Ponty, M. (1963).

8 I had to wonder whether I should be working at being 'grumpy' to gain Hugh's acceptance rather than forcing a strained smile when I was feeling frightened or just mildly anxious.

9 Networks of course that a researcher has to follow as well to try and map their links and understand how dispositions and competencies can move an actor through them.

10 Formal training is given on-shore before boarding the boat and during expeditions on sail training boats, but as Hutchins' (1996) research demonstrates, much of the practical knowledge has to be learnt *in action*. It can be anticipated verbally, visually and emotionally and reflected upon *post hoc*, yet the moment of situated *doing* is the key.

11 See Du Gay (1996) and Hughes (forthcoming) on the analysis of the folding of consumption practices back into production practices.

12 Geertz (1983: 167) 'Like sailing, gardening, politics and poetry, law and ethnography are crafts of place: they work by the light of local knowledge.' Dee and Hugh knew of these places because like ethnographers they had 'been there'. These were places where modernity de-intensified, unfolded, slowed down – it still met their 'secret coves' but in a glancing curve. Dropping a few rusty bits of metal on the beach, alongside orange plastic floats from trawler nets, some sheep by way of European subsidy that kept the trees down. And us of course in search of shelter and a 'temporary address' (Grossberg 1992).

13 They may well be socially reproduced informal systems as argued earlier, yet they have the effect of creating a sense of authorship in Hugh and thereby a unifying sense of agency.

14 Scottish shipbuilding was in decline during the entirety of Hugh's career.

15 It is my hope that readers of this chapter will have been alerted to my own narrow focus on Hugh's emotions which was how it happened in the field so to speak. Dee's skillful hosting actually made her work less visible, to my eyes at least and made my difficulties with Hugh all the more apparent.

References

Averill, J. R. (1974) An analysis of psychophysiological symbolism and its influence on theories of emotion, *The Journal for the Theory of Social Behaviour*, 4: 147–90.

Bachelard, G. (1964) *The Poetics of Space*, (trans. Maria Jolas), Orion Press, New York.

Bourdieu, P. (1977) *Outline of a Theory of Practice*, Cambridge University Press, Cambridge.

Bourdieu, P. (1984) *Distinction: a social critique of the judgement of taste*, (trans. R. Nice), Harvard University Press, Cambridge MA.

Bourdieu, P. (1990) *The Logic of Practice*, Polity Press, Cambridge.

Boyer, C. M. (1994) *The City of Collective Memory*, MIT Press, London.

Cohen, S. and Taylor, L. (1992) *Escape Attempts: the theory and practice of resistance in everyday life*, Routledge, London.

Cook, I. and Crang, M. (1995) *Doing Ethnographies*, IBG Catmog Series, London.

Cook, I. (1995) A Grumpy Thesis : geography, autobiography, pedagogy, unpublished PhD: University of Bristol.

Du Gay, P. (1996) *Consumption and Identity at Work*, Sage, London.

Euromonitor (1991) *Market Research GB*, 9.

Geertz, C. (1983) *Local Knowledge*, Basic Books, New York.

Gieryn, T. (1983) Boundary-work and the demarcation of science from non-science: strains and interests in professional ideologies of scientists, *American Sociological Review*, 48: 781–93.

Goffman, E. (1967) *Interaction Ritual*, Pantheon, New York.

Grossberg, L. (1992) *We Gotta Get out of this Place: rock music and conservatism*, Routledge, London.

Hincheliffe, S. (forthcoming) Home-made space and the will to disconnect, in *Ideas of Difference: social space and the labour of division* (ed. Hetherington, K. and Munro, R.), Blackwell, Oxford.

Hochschild, A. (1983) *The Managed Heart: the commercialisation of human feeling*, University of California Press, Berkeley.

Hughes, A. (forthcoming) Constructing competitive spaces: on the corporate practice of British retailer–supplier relationships, *Environment and Planning A*.

Hutchins, E. (1996) *Cognition in the Wild*, MIT, Boston.

Kleinman, S. and Copp, M. (1993) *Emotions and Fieldwork*, Sage, London.

Kraus, R. (1987) *Recreation and Leisure in Modern Society*, Scott Foresman, Glenview, IL.

Krouwel, B. and Goodwill, S. (1994) *Management Development Outdoors*, Kogan Page, London.

Lamont, M. (1992) *Money, Morals and Manners: the culture of the French and the American upper-middle class*, University of Chicago Press, London.

Latour, B. (1988) *Science in Action*, Open University Press, Milton Keynes.

Merleau-Ponty, M. (1963) *The Structure of Behaviour*, Beacon Press, Boston.

Mintel (1997) *Mintel Leisure Intelligence*, February.

Morris, R. C. (1995) All made up: performance theory and the new anthropology of sex and gender, *Annual Review of Anthropology*, 24: 567–92.

Parr, H. (1998) Mental health, ethnography and the body, *Area* 30(1): 28–37.

Parrott, G.W. and Harré, R. (1996) Embarrassment and the threat to character, in *The Emotions, Social, Cultural and Biological Dimensions* (eds Parrott, G.W. and Harré, R.), Sage, London, pp. 39–56.

Parry, B. (forthcoming) *Hunting the Gene Hunters: the role of hybrid networks, status and chance in conceptualising and accessing corporate elites*, Environment and Planning A.

Purves, L. (1996a) On cruising, *Yachting Monthly*, January: p. 59.

Purves, L. (1996b) On cruising, *Yachting Monthly*, July: p. 65.

Rojek, C. (1985) *Capitalism and Leisure Theory*, Tavistock, London.

Rojek, C. (1995) *Decentring Leisure: rethinking leisure theory*, Sage, London.

Sekla, A. (1996) *Fish Story*, Richter Verlag, Rotterdam.

Shields, R. (1991) *Places on the Margin*, Routledge, London.

Skeggs, B. (1997) *Formations of Class and Gender*, Sage, London.

Veblen, T. (1925) *The Theory of the Leisure Class*, Allen & Unwin, London.

Willis, P. E. (1977) *Learning to Labour: how working class kids get working class jobs*, Saxon House, Farnborough.

Wolf, M. (1992) *A Thrice-Told Tale: feminism, postmodernism, and ethnographic responsibility,* Stanford University Press, Stanford CA.

Wynne, A. (1988) Accounting for accounts of the diagnosis of multiple sclerosis, in *Knowledge and Reflexivity: new frontiers,* (ed. Woolgar, S.), Sage, London, pp. 101–22.

15

LEISURE PLACES AND MODERNITY

The use and meaning of recreational cottages in Norway and the USA

Daniel R. Williams and Bjørn P. Kaltenborn

Introduction

When we think of tourism we often think of travel to exotic destinations, but modernization has also dispersed and extended our network of relatives, friends, and acquaintances. Fewer people live out their lives in a single place or even a single region of their natal country. Modern forms of dwelling, working, and playing involve circulating through a geographically extended network of social relations and a multiplicity of widely dispersed places and regions. Much of the "postmodern" discourse on tourism leaves the impression that tourists seek out only the exotic, authentic "other" and experience every destination through a detached "gaze" that rarely engages the "real" (i.e., uncommodified) aspects of the place (MacCannell 1992, Selwyn 1996, Urry 1990). Contrary to images of "gazing" tourists on a pilgrimage for the authentic, much of modern tourism is rather ordinary and involves complex patterns of social and spatial interaction that cannot be neatly reduced to a shallow detached relation. Leisure/tourism is often less packaged, commodified, and colonial than contemporary academic renderings seem to permit.

One widespread, but largely unexamined form of leisure travel involves the seemingly enigmatic practice of establishing and maintaining a second home (what we will generally refer to as *cottaging*). Like the tourist, the cottager often appears to be engaged in an effort to reclaim a sense of authentic place or identity that otherwise seems so elusive in the fragmented postmodern world (MacCannell 1976). But it is not the nomadic and ongoing pilgrimage that often character-izes tourism. Rather the very word *home* implies becoming native to a place, setting down roots, and investing oneself in a place. Because cottagers are simultaneously tourists and residents of their cottage locale, leisure related to cottaging represents a unique context within which people encounter, come to know, and transform places.

Given the goal of this volume to broaden the description of leisure/tourism practices by examining them as vehicles for attaining geographic knowledge, our purpose is to examine the leisure practice of second home use and ownership. Maintaining a second home is common in many countries (Coppock 1977). Although cottaging shares an emphasis on leisure across cultures, it is neverthe-less embedded in varying social histories and traditions of use and meaning. We examine this leisure practice by contrasting cottage use in Norway and the USA – two nations representing different histories, identities, societal development, and outdoor use traditions. Specifically, we examine cottaging within the context of two important and related geographic themes: (1) how cottage use relates to the identity dilemmas created by modernity and the phenomena of space-time compression; and (2) what cottaging tells us about sense of place as a form of geographic awareness. We begin with some background on modernity and its impacts on identity and sense of place to contextualize our two case studies of cottage use.

Modernity, identity and a sense of place

Modernity possesses an excess of meaning and no meaning at all: "The fore-boding generated out of the sense of social space imploding in upon us . . . translates into a crisis of identity. Who are we and what space/place do we belong?" (Harvey 1996: 246). This problem is also reflected in research on com-munity, home, migration, and tourism, which are infused (and encumbered) by outdated assumptions of a geographically rooted subject. The movement of peoples, rather than being seen as an integral aspect of social life, "has been regarded as a special and temporary phenomenon which has been examined under the headings of migration, refugee studies, and tourism" (Hastrup and Olwig 1997: 6). With circulation and movement more the rule than the excep-tion an important geographic dimension of leisure practices is to understand how people in differing cultural contexts use leisure and travel to establish identity, give meaning to their lives, and connect with place.

Modernity is, in large part, experienced as a tension between the freedom and burden to fashion an identity for oneself. Aptly described by Marshall Berman in *All That is Solid Melts into Air: The Experience of Modernity*, this tension originates in:

> our desire to be rooted in a stable and coherent personal and social past,
> and our insatiable desire for [economic, experiential, and intellectual]
> growth that destroys both the physical and social landscapes of our past,

and our emotional links with those lost worlds; our desperate allegiances to ethnic, national, class, and sexual groups which we hope will give us a firm "identity," and the internationalization of everyday life – of our clothes and household goods, our books and music, our ideas and fantasies – that spreads all our identities all over the map; our desire for clear and solid values to live by, and our desire to embrace the limitless possibilities of modern life and experience that obliterate all values.

(Cited in Sack 1992: 6)

With the melting (and profaning) of traditional sources of meaning, leisure stands to fill the void. Yet consumerism and mass tourism, as forms of leisure, are also subject to many of the same forces that contribute to the dilemma of meaning (Rojek 1995, Sack 1992). Leisure represents the increased freedom or capacity to seek out and express identity and the burden of discovering meaning in a meaningless world. Guided by fewer strictures on how and what to choose, the self is at once liberated through the expansion of leisure and, at the same time, saddled with the burden of making choices from among an ever widening market of options. Ironically, for all the choice and freedom to construct the self (i.e., the freedom of leisure), our personal appropriation of life choices and meanings is often constrained by highly standardized market-driven modes of production and consumption. Consequently, leisure and tourism may be experienced as potentially authentic, personalized, and identity-enhancing or increasingly manufactured, commodified, and disorienting.

An important geographic feature of modernity that contributes directly to both the freedom/burden of identity and the expansion of leisure/tourism is the changing nature of time-space relations. The experience of modernity is associated with rapidly accelerating rates of exchange, movement, and communication across space. As Massey (1993) points out, modern experiences such as time zones, jet lag, future shock, bank cards, ATMs, fax machines, e-mail, the Internet, and the global village are all symptomatic of time-space "compression." Modernity "breaks down the protective framework of small community and of tradition, replacing these with much larger, impersonal organizations. The individual feels bereft and alone in a world in which she or he lacks the psychological supports and sense of security provided by more traditional settings" (Giddens 1991: 33). With each new symbolic connection to the larger world "the traditional face-to-face community loses its coherence and its significance in the life of its participants. . . . Their sense of 'belonging' is no longer only, or even primarily, rooted in the local soil" (Gergen 1985: 215). The resulting crisis of meaning leads inevitably to arguments that "in the middle of all this flux, one desperately needs a bit of peace and quiet; and "place" is posed as a source of stability and an unproblematic identity" (Massey 1993: 63).

Though there is widespread sentiment that the insecurity associated with space-time compression leads to the search for authentic and stable place, the social or moral value of maintaining a unitary sense of place or strong place-bonds is not beyond reproach. On the one side, certain anti-modernist environmental and social philosophies (e.g., bioregionalism and communit-arianism) suggest a return to a mode of "dwelling in place" that is thought to have been more common in past centuries. They suggest that by sinking roots more deeply into a place we can "prevent the place from eroding into McWorld – from becoming lost in the abstract space of global markets" (Sagoff 1992: 369).

Other critics see sense of place as a romantic, dangerous sentiment to re-create Tönnies' *Gemeinschaft* (Shurmer-Smith and Hannam 1994). They fear that bioregionalist and communitarian politics hold the potential to be "exclusionary" (i.e., racist and indifferent to distant "others"). "The danger arises when such modes of thought are postulated as the sole basis of politics (in which case they become inward-looking, exclusionary, and even neo-fascistic)" (Harvey 1996: 199). In this view it is naïve to believe that such societies will respect the pos-itive enlightenment values of human diversity and justice for the "other."

Massey (1993) argues for a "progressive" view of place that tries to give cred-ibility to the human need for authenticity and rootedness, recognizing that such sentiments need not be construed as a case for exclusionary communities. "The question is how to hold on to that notion of spatial difference, of uniqueness, even of rootedness if people want that, without it being reactionary" (Massey 1993: 64). Thus "what gives a place its specificity is not some long internalized history but the fact that it is constructed out of a particular constellation of rela-tions articulated together at a particular locus" (Massey 1993: 66). Places constitute unbounded constellations of global and local process. Modernity makes this wider constellation of relations possible, but doesn't necessarily reduce place uniqueness. Thus, a place may have "a character of its own," but it is still possible to feel it without subscribing to the Heideggerian notions of essentialism and exclusivity. Though place identities often lack the singular, seamless, and coherent qualities frequently attributed to the idea of sense of place, multiple place iden-tities can be, and often are, both a source of richness and conflict.

Similarly, Sack (1992, 1997) argues that the condition of modernity is not so much about a decline or loss of place-based meaning, as it is often interpreted, but about a change in how meaning is created or constituted in the modern age. These processes thin the meaning of places. "From the fewer, more local, and thicker places of premodern society, we now live among the innumerable inter-connected thinner places and even empty ones" (Sack 1997: 9). In this context leisure/tourism represents a major complicitous factor of transforming places *and* the thinning of meaning, the epitome of "consuming places" (Urry 1996). This makes the very authenticity many tourists seek increasingly inaccessible

(MacCannell 1976). How much and under what circumstances can tourism balance the inevitable tension between the commodification (thinning) of places and our desire to experience and live in unique and "thickly" textured places?

Case studies in cottage culture

Building on Sack's distinction between thick and thin places and Massey's notion of a progressive sense of place, we examine the use and meaning of recreational second homes through two case studies, one in northern Wisconsin and one in southeast Norway. Their comparison provides a unique way to observe how modern place-based identities are forged within a wider assortment of local and global interactions. The contexts of seasonal migration and tourism also empha-size the ways in which the local is partially constituted from the global (or at least the larger regional) in the form of various urban migrations associated with tourism across the seasons.

The northern Wisconsin site, known locally as "Hayward Lakes", is in a region commonly identified by midwesterners as the "Northwoods" to differentiate it from the more densely populated prairie to the south. It has rural character marked by diversity of land ownership, including seasonally occupied lakeside homes (50 per cent of the dwellings in the area), resorts, and campgrounds, and large tracts of county, state, and federal forest land. The area is also adjacent to the Lac Courte Oreilles Indian Reservation which operates a modest casino on the outskirts of Hayward (population 1,900), the major commercial center in the area. The surrounding landscape is dominated by forest with dozens of inter-spersed lakes that provide cottage sites and many recreation activities (e.g., fishing, hunting, boating, skiing, snowmobiling, and mountain biking). Several "event" attractions that bring thousands of tourists to the region have evolved from these recreation activities, e.g., the American Birkebeiner is modeled after the famous Norwegian race by the same name and is the largest cross-country ski race in North America. There is strong regional influence of nineteenth-century Scandinavian immigration. Constituting an amalgamation of traditional-rural and modern-urban lifestyles, the region has experienced rapid transformation as a tourist destination in recent years. The majority of cottage owners and tourists originate from the nearby urban center of Minneapolis-St. Paul, Minnesota (225 km southwest) or Chicago, Illinois (650 km southeast).

Recreation and tourism in the area prior to World War II was largely restricted to rustic summer resorts, cottaging, and traditional activities (i.e., fishing and hunting) that followed earlier logging. Tourism/leisure has dramatically changed from the dominance of fishing and most family owned and operated lakeside resorts have been replaced by seasonal homes. New recreational technologies, snowmobile and mountain bike, have impacted and, in conjunction with skiing, expanded the tourism season into the fall and winter. With limited additional waterfront home sites, prices have risen dramatically in recent years.

The Norwegian site, known as Sjodalen, is a broad valley located in a rugged alpine and sub-alpine region at the outskirts of Jotunheimen (Home of Giants) National Park, a rural part of Norway four hours' drive (200–400 km) from urban centers Olso and Bergen. The valley holds trails, roads and buildings associated with long subsistence use. Most of the land in the study site is state owned common property. The valley surroundings are diverse, with highland pine and birch forests, non-forested alpine lands, enclaves of pastures and, traditionally, cultivated fields around the old summer farms ("seters"); lakes, streams and a world-class rafting river. The area attracts people from urban regions and nearby communities, with a range of recreational opportunities; hiking, cross-country and downhill skiing, fishing, hunting. Although Sjodalen is primarily a recreational and second home area, it was previously used for logging and there are still summer farms scattered throughout the area; some in traditional use, but many converted into cottages, or neglected and thus fallen into ruin over time.

There are few permanent residents, unlike the Wisconsin site. Most are associated with the tourism industry and the handful of small resorts throughout the valley. It has provided a corridor for travel between the western and eastern part of Southern Norway, and today a highway bisects the area. The area is extremely

Plate 6 Summer farm, Sjodalen

Source: Bjørn P. Kaltenborn

219

popular and there is a great demand for more second homes. Present land-use zoning limits the potential for additional cottages in the area and consequently prices are increasing.

Escaping modernity: from thin to thick

As a reaction to the modern tendency to experience thinned out places, cottaging can represent an effort to construct an identity firmly anchored in place. Building on interviews and comments of cottage owners we discuss three themes: back-to-nature, inversion, and continuity as meanings of second homes (Jaakson 1986). In our adaptation of these categories each is embedded in a broader framework of modernity, identity, and sense of place; that is, each reflects the tension between the tendency to thin the meaning of places and to establish a sense of home and identity.

Back-to-nature

The most direct form of escape from modernity is to seek refuge in nature, (Jaakson op. cit., Halseth 1992). The setting of the cottage affords greater access to nature, in typically rural locations (i.e., "Northwoods" or "in the mountains"), and by virtue of the relative spaciousness of most cottage developments. Relaxing around the cottage combines with experiencing nature through walks, picnics, viewing wildlife, fishing, canoeing and gardening (Stynes et al. 1997). Among Hayward Lakes cottage owners there is far more emphasis on passive nature appreciation than more active forms of leisure such as golf, tennis, and water-skiing. Cottage use averages 72 days increasingly spread through the year. One woman cottager describes back-to-nature as "stepping back in time," "a sweet-ness and a simplicity to it," "rustic, but not camping out," "unsophisticated," "unadorned" and "an unadulterated environment." One male informant describes it as a quality of consciousness more than of natural surroundings:

> It's just a totally different feel, I mean it's the woods, the trees, the lakes, the water, wildlife, birds, sea gulls go by, yeah so I . . . think you see things and feel things differently when you're here versus the city. Its not that there aren't birds and trees in the city, it's just that you have a different focus. You're working . . . where up here it's just the opposite. You're into relaxing and getting away from everything.

A major role of the Norwegian cottage is to provide immediate contact with nature. Most cottages in Sjodalen have natural surroundings with attractive views and some privacy from neighbors. The topography permits placing cottages so that they seldom crowd one another. The average user spends six weeks a year at the cottage, an average of eight days during harsh midwinter. Activity levels vary considerably, typically short and long hikes, skiing, fishing and hunting, picking berries and mushrooms. Most cottagers spend a great deal of time at the cottage just relaxing and/or engaging in practical activities. Traditional and moderately strenuous activities like fetching water, gathering and heating with firewood, and maintaining the building, bring people in contact with their immediate surroundings and provide elements of well-being and give meaning to cottage life. Many cottages have no electricity or running water, many have road access only during the summer season. The lack of modern facilities seems to be a conscious choice of most of the cottage owners.

Many cottage owners cite a lack of modern facilities as important to facilitate nature experience, the feeling of well-being at the cottage largely a function of being comfortable with simple means. Such a feeling is also evident in the importance placed on adjusting lifestyles more to natural daylight and darkness, escaping the sounds of civilization, experiencing distinct differences in weather. "A lot is important, particularly getting close to nature, feeling the wind and weather, having a few people around you, and living as one with nature and the rhythms there" (middle-aged man). "You become particularly fond of special mountain peaks. To watch these at all hours of the day – in all kinds of weather is a gift" (retired man). "Nature" becomes a metaphor for empowerment, identity, and a particular experience of embodiment.

Inversion

A most pervasive meaning of a second home is that "life at the cottage is lived differently," an "inversion" (Jaakson 1986: 377), a vacation from city daily life. The cottage is a center for leisure, the primary home a center for work. The most strongly endorsed statement in a random sample of Hayward Lakes cottagers was that life was definitely "less hectic," "very different." A twelve-year veteran of the area speaks for many cottagers:

> It is just great and the people up here are relaxed. They're not high tensioned, high stressed like you find in the city or work. So you forget about all that. You almost forget about what day it is.

> Like 100 percent different (Chicago woman, laughing).

A particular way in which life is lived differently at the cottage is the convergence of work and leisure. Here leisure is not divorced from the rhythms of life as in the "work home." With the cottage as the setting or center of all activities, doing chores, maintaining property and landscape, recreating, being outside, visiting with the family, the line between work and leisure fades. Maintenance of the property is seen as integral to life at the cottage. One man who hired designers for the interior and garden of the cottage did no first-home yard work, but claims to spend the bulk of his time at the cottage doing so.

The relaxed pace of life is also linked to the different way work is perceived. One informant, a geologist for an environmental consulting firm, admits to bringing some work to the cottage, but describes cottaging as a different frame of mind:

> for me the switch coming up here is that down there [meaning a community 30 miles north of Chicago] every minute at work counts. You're worrying about it, you're [thinking] "what to do this minute, what do I do the next minute" and then the next day. I have to fly down to Chicago for a meeting for one day and then back up again. It's happened a couple of times before. But mostly I try to set it up where people know not to bother me unless it's an absolute emergency.

This does not necessarily mean that people are bringing the office to the cottage. Many speak of the transition to the Northwoods as leaving the world of work behind. One informant describes it as a kind of "rinsing out work" from his mind that occurs after about an hour's drive from the city.

Inversion in the Norwegian case can be more pronounced among the Norwegian cottage owners. In interviews and surveys, the cottage owners repeatedly emphasize the "different-ness" of the life at the cabin. It is clearly a retreat from everyday life, and a rest from the pressures from modern life:

> [I] feel comfortable in my own company. The cabin and the mountains give me peace of mind. I often bring with me things to the cabin which I don't feel I have time to deal with at home. At the cabin I can permit myself just "to be" without doing anything. Just sit and enjoy the view.

This does not mean being entirely passive and avoiding work altogether. The typical cottage owner in Sjodalen is less reliant on modern technologies and actively applies practical skills for maintaining the place and harvesting resources like berries, mushrooms, and fish:

> [It's] another environment, without all the technical appliances, nice not to have all the electrical equipment, use candle lights – wood heating, all in all live a simpler life.

Most of this is carried out at a relaxed pace without too much planning and little or no pressure to get finished at any particular time. Considerable amounts of time are spent "doing nothing in particular" (i.e. reflecting on this and that without being too concerned about time). Overt activities like small building repairs or cutting firewood interchange with spells of contemplation throughout the day. Such contemplative interludes are salient in the broader picture of inversion. Reflections about one's own existence, social life, and activities are often interpreted within the context of the natural surroundings. The central meaning of the cottage life is created through opportunities to live a comfortable and relaxed existence, with a feeling of control over one's life in attractive surroundings, without having to work too hard for it.

Continuity and sense of place

Tuan (1977) describes *place* as a center of meaning. In a globalized world many feel placeless, the cottage is a center of meaning across the life course even as the permanent residence changes. The cottage provides continuity of identity and sense of place through symbolic, territorial identification with an emotional home. In the Hayward Lakes region territorial identification occurs on at least three levels of scale: the cottage itself and its immediate surroundings as a sense of home; the lake as a social system, as neighborhood or community; and as regional landscape of shared cultural values.

The cottage provides continuity across generations within the family and across the life course within a generation. The cottage provides for family togetherness that is distinct from often segmented lives and schedules that are felt to characterize most urban households (Jaakson 1986). Daily paths and projects of individual family members are spatially bounded and more interwoven at the cottage, a site of family memories, providing symbolic territorial identification across generations, typically expressed:

> My grandfather built it in the 30's. This whole area had been logged and he was the original owner of the Edgewater Lodge. It's still in my father's name, but my brothers and I come up here and take care of it too. [We are] trying to make sure that it stays in the family . . . all six of us brothers still use it. And my folks come out here two to two and a half months a year.

When asked about future plans for the place, responses almost always include a reference to passing the place on to the children. Jaakson (1986) notes that this is rarely the case for the primary home. He describes a case of two college-aged brothers whose parents had died:

They promptly sold the urban house where they had lived since childhood. But when asked if they would also sell the cottage, they looked
aghast and replied: "We'd *never* sell the cottage!" . . . The cottage was
their emotional *home*, the city house a mere residence.

(Jaakson 1986: 381).

Many respondents express place-identity and continuity as reasons they have a
cottage:

At that time [early 1980s], we paid $52,000 for this land and 250 feet [of
shoreline] and nobody on this lake had ever paid that much for land, so we
were laughed at. People thought we were crazy, but we were buying not
only a piece of property, we were buying our dream. I don't know how to
explain what it is about this place but there is a sense of being home.

We moved and owned 8 to 10 different homes and so we've never had
any sense of ownership that was worthwhile to have, anything that was
going to be there.

Shared identification with neighbors (particularly other property owners around
the lake) is important, people sought a cottage amongst "community spirit" and
"where everyone knows one another." Most Wisconsin respondents, especially
those with longer tenures, discuss neighbors by name and describe social practices such as "puttsing" around the lake visiting neighbors and hanging out at
taverns and restaurants located around the lake. A number of respondents
suggested that they have more friends and more social life at the cottage than
they do at their work home. This was very different among the Norwegian cottage
owners who focus on family and friends who accompany them at the cottage,
with little socializing amongst the cottage owners. The cottage is primarily a
place for maintaining familiar social bonds.

Another symbolic identification has to do with the mystique of the "Northwoods," a symbolic place routinely reproduced in the public discourse of what
it means to live in Wisconsin and neighboring states. Among residents of
the upper Midwest (northern Illinois, Wisconsin, Michigan, Minnesota) the
"Northwoods" and "up north" refer to a distinct place that is symbolically "a part
of the ritual retreat from the city," "(t)o most, the allure of the Northwoods is
that it simply is what the city is not: pristine, wild, unspoiled and simple."
(Bawden 1997: 451). This is a pristine myth (the Northwoods were "cutover"
100 years ago); nevertheless, the area is still "quite different from the cities,
towns, or dairyland countryside to the south" (op. cit.: 451). Cottagers clearly
experience this as they make their trek north, noting changing landscape, the

composition of the forest ("all of a sudden the white birches become the predominant tree in the forest") and cultural elements, like a style of barn. Many often name specific towns that mark the transition. In one small Wisconsin town there is a monument on a large rock which proclaims to mark the spot as "half way north," referring to its location at latitude 45 degrees north of the equator.

Sjodalen continuity involves identification with an emotional home, though not necessarily of childhood vacation experiences, but imbued with an outdoor ethic:

> Cabin life is tradition. It has always been my favorite existence. Away from crowds, time pressures and the regulations of civilization and technical things. Experience, enjoy and protect "untrammeled" nature. Live a simple life materially and in agreement with a few, chosen people. Physical – and not the least psychological – balance is quickly restored during a stay at the cabin.
>
> (middle-aged woman)

A common theme across the Nordic countries is that people often seek cottages in the same district of their family origin (Nordin 1993), and a search for continuity is prevalent for a great portion of the cottage owners in Sjodalen. Continuity is emphasized through the importance of certain traditions associated with cottages. To some the annual ptarmigan hunt (a type of mountain grouse) or trout fishing become ritualised amongst particular friends. Less specific, but perhaps even more important traditions, include spending Easter and Christmas holidays at the cottage.

At a cultural or national level, the significance of cottaging for Norwegians is reflected in the very high levels of recreational homes per capita. In both Norway and Sweden there are between 70 and 80 second homes per 1000 population compared to 20 to 30 per 1000 in the Great Lake US states. Explanations have been offered for the popularity of seasonal homes in Scandinavia (Nordin 1993), predominantly the relatively recent depopulation of the countryside. People have a great desire to own a cottage in the district of forebears. In Norway, cottage use is also linked to summer farms where people moved up into farmhouses near the higher pastures to tend livestock. This made for a pleasant break in the routine. However, family heritage is likely only to be part of the explanation for popularity in Scandinavia. As a project to seek meaning and create and maintain identity, the diverse roots of "cottage culture" in Scandinavia are important. Cottaging bridges different epochs and domains of societal development.

Historically, extensive use of remote resources and outlying areas created a need for small cabins and summer farms. Many were later converted to recreational homes but have their roots in rural subsistence. Almost without exception these are still owned by people in rural districts or by the state. Many rural people today still build cabins close to their primary residence, an attachment to place and heritage more salient in choice of cabin location than it is for urbanites.

At the turn of the century two additional cottage trends emerged. Individuals from the urban upper class, notably in England but also to some extent in Norway, erected exuberant retreats in the countryside, particularly in order to allow extended salmon fishing and large game hunting. Second, larger industrial companies established small "cottage communities" near the cities for the use by their employees. This working class cottage culture has since dissolved as most of the cottage communities have been sold off to individuals (although many companies still own cottages intended for more executive use). In social-democratic Scandinavia, however, the working class leisure ideal, which included the notion of residing in nature during leisure, diffused into the middle classes during the prosperous 1960s and 1970s. During this time many locations in the rural districts were opened up and even developed for recreational housing, allowing cottages to become a widely shared practice in Norwegian life.

In comparing the two case studies, cottage culture in Norway clearly involves a strong component of reinforcing national identity. That is, Norwegian use of cottages is very much embedded in "Norwegian-ness." In an effort to fortify the sense of nationhood such "typically Norwegian activities" as seeking "peace and quiet," "going for a walk," and cross-country skiing become symbols of national identity (Erikson 1997). They constitute focal symbols in a national debate about globalization and European unification (Kaltenborn et al. 1993). As leisure practices, these symbols are highly ritualized through cottage use – particularly as it reproduces traditional subsistence activities (e.g., hunting, fishing, mushroom and berry picking) and modes of dwelling (e.g., wood fires, fetching water, natural light, etc.). The cottage serves as a center of peace and quiet and a base for nature outings. This is diversifying, however, as cottage culture is modernized. The development of modern, fully equipped cottages may be attracting new, different groups, creating conflicting social norms. The modernization and commodification of cottaging unsettles Norwegian cottage identity.

In contrast, Wisconsin cottaging has a strong regional significance shared among residents of the upper Midwest as a "Northwoods" or "up-north" culture. Northwoods cottage use traces its origins back to a lakeside fishing and/or hunting cabin, many converted from small vacation resort properties. The historical tie to non-leisure uses, ancestral homes and village is less strong here than in the Norwegian case. Though Americans possess myths related to the conquest of nature, wilderness, and the frontier, the American identification with nature is not wrapped in cottage use to the same degree as it can define the Norwegian character. The upper Midwest emphasis on the summer cottage lifestyle may have some origins in the dominance of Scandinavian immigration and settlement of

Minnesota and Wisconsin. It may also be linked to the common experience of northern latitudes (e.g., the common "feel" of northern boreal forests and the emphasis on summer as a leisure time and place).

Conclusions: dilemmas in the creation of meaning

An attachment to place or community, the sense of "insidedness" made possible by a lifelong accumulation of experiences in place, is important in maintaining a sense of personal identity (Giddens 1991, McHugh and Mings 1996). Yet the modern identity is no longer firmly rooted in a singular local place. The leisure practice of cottaging (owning and using a second home) embodies both sides of this modern tension. It is a modern expression of the need to have an authentic, rooted identity somewhere, but also a concrete manifestation of a segmented, isolated self-living in more than one place. On the one side, cottaging represents an escape *from* modernity and *towards* a sense of place, rootedness, identity, and authenticity, echoing Crouch (1994: 96): "escape becomes an escape for home, not just from home." Cottaging is an attempt to thicken the meanings we associate with places in response to the modern tendency for places to become thinned out. Cottaging inverts much that is modern against the modern tendency to separate and segment. It emphasizes continuity of time and place, a return to nature, and con-vergence of spheres of life such as work and leisure. However, cottaging is very much an extension of modernity. Cottage use is motivated by and played out in the modern context of globalized cultural production and accelerated time-space relations. It necessarily re-creates the segmented quality of modern identities. It does so in the form of separate places for organising distinct aspects of a fragmented identity. It narrows and thins out the meaning of each "home" by focusing the meaning of each on a particular segment of life (i.e., work and subsistence of urban daily life versus recreation and rejuvenation of cottage life). It also segments identity around phases in the life cycle with youth and retirement focused more on cottage life than working adulthood.

Modern forms of dwelling, working, and leisure involve circulating through a geographically extended network of social relations and a multiplicity of widely dispersed geographic places. Circulation no longer represents an interruption of ordinary, settled life, but constitutes a normal condition for many people (Olwig 1997). Given contemporary unsettled life, it should not seem unusual to set down multiple roots in multiple places.

Recognizing that circulation and travel are increasingly the norm effectively deterritorializes or *dis*-places what have long been geographically bounded conceptions of culture, home, and identity and makes problematic certain salient themes in the study of tourism. The discussion of many interesting and familiar tourism practices is virtually absent in the postmodern discourse on tourism.

These include, for example, visiting friends and family, seasonal migrations linked to cycles of work and leisure, charitable uses of leisure travel (i.e., building homes for "Habitat for Humanity" or a church project to build a community school in rural Mexico). Numerous surveys of tourist patterns suggest that a significant volume of tourism involves visiting extended family members, old friends who have moved, and new acquaintances made through spatially extended social networks (Pearce 1982). These forms of leisure circulation have implications for geographical knowledge.

The second theme concerns center-periphery relations in tourism. Tourism flows are generally conceptualized as moving from the urban center to the rural periphery. In our case studies, however, many cottage owners have deeper roots in the cottage region, with ownership stretching back several generations, than in what geographers would call their primary home. In terms of time, money, and psychic energy, people invest considerable portions of their lives in these "secondary" places. To necessarily privilege one (usually the urban work place) as primary, central, and everyday and the other as secondary, peripheral, and exotic seems unwarranted.

The third theme pertains to the relation of the tourist to the site as captured in notions of the tourist gaze and authenticity. The questions become refigured. Which place embodies "ordinary" life and which place yields the "extra-ordinary" life? Which place is authentic and genuine and which is fabricated and artificial, or how does the subject feel "ownership" and identity? The tourism literature typically presumes that the anchor for normalcy, originality, and ordinariness is daily urban life and that vacations, travel, and tourism refer to seeking or gazing upon the extra-ordinary, the exotic, and *otherness*. The idea of the tourist gaze implies a contrast between the ordinary and the extraordinary with the tourist experience primarily one of looking upon "features of landscape and townscape which separate them off from everyday experience" (Urry 1990: 3). When movement and circulation are the norm rather than the exception, when people regularly reside in more than one place, it is difficult to say which is the everyday place and which is the extra-ordinary one.

Unlike many forms of tourism and holiday making, with cottaging there is little sense of busy-ness, no rush to take it all in. There are no schedules to co-ordinate, no tours to arrange, no guidebooks listing operating hours to consult, no concert tickets to buy, no hotel concierge, and no tour bus. At the cottage one can stop consuming, one can stop "collecting signs" for a while. What is ordinary is the second home, the natural landscape and scenery: it is the modern, urban life that is felt extra-ordinary. For many the cottage is a literal and emotional home. In the modern world the identity we associate with home is increasingly created through residence in more than one place.

Acknowledgements

The authors wish to acknowledge the financial support of the US Department of Agriculture, North Central Forest Experiment Station and the Aldo Leopold Wilderness Research Institute, as well as the Norwegian Research Council's program on Environmentally Dependent Quality of Life. We also wish to thank Mike Maehr, Norm McIntyre, Susan Van Patten, Jarkko Saarinen, Susan Stewart, and Alan Watson for their contributions to this effort.

References

Bawden, T. (1997) "The Northwoods," in R. C. Ostergren and T. R. Vale (eds) *Wisconsin Land and Life,* University of Wisconsin Press: Madison.

Coppock, J. T. (1977) *Second Homes: Curse or Blessing?* Pergamon: London.

Crouch, D. (1994) "Home, escape and identity: rural cultures and sustainable tourism," in B. Bramwell and B. Lane (eds) *Rural Tourism and Sustainable Rural Development*, Channel View Publications: Clevedon.

Erikson, T. (1997) "The nation as a human being – a metaphor in a mid-life crisis? Notes on the imminent collapse of Norwegian national identity," in K. F. Olwig and K. Hastrup (eds) *Siting Culture: The Shifting Anthropological Object*, Routledge: London.

Gergen, K. J. (1985) *The Saturated Self: Dilemmas of Identity in Contemporary Life*, Basic Books: New York.

Giddens, A. (1991) *Modernity and Self-Identity*, Stanford University Press: Stanford.

Halseth, G. (1992) "Cottage property ownership: interpreting spatial patterns in an eastern Ontario case study," *Ontario Geography* 38: 32–42.

Halseth, G. and Rosenberg, M. W. (1995) "Cottagers in an urban field," *Professional Geographer* 47: 148–59.

Harvey, D. (1996) *Justice, Nature and the Geography of Difference*, Blackwell: London.

Hastrup, K. and Olwig, K. F. (1997) "Introduction," in K. F. Olwig and K Hastrup (eds), *Siting Culture: The Shifting Anthropological Object*, Routledge: London.

Jaakson, R. (1986) "Second-home domestic tourism," *Annals of Tourism Research* 13: 367–91.

Kaltenborn, B., Gøncz, G., and Vistad, O. I. (1993) "På tur i felleskapet: mulige virkninger av EØS og EU på den norske allemannsretten" (A trip in the commons: possible impacts of the European Economic Agreement and the European Union on the Norwegian common access tradition), Lillehammer: Østlandsforskning.

MacCannell, D. (1976) *The Tourist: A New Theory of the Leisure Class*, Schocken: New York.

MacCannell, D. (1992) *Empty Meeting Grounds: The Tourist Papers*, Routledge: London.

McHugh, K. E. and Mings, R. C. (1996) "The circle of migration: attachment to place in aging," *Annals of the Association of American Geographers* 86: 530–50.

McHugh, K. E., Hogan, T. D., Happel, S. K. (1995) "Multiple residence and cyclical migration: a life course perspective," *Professional Geographer* 47: 251–67.

Massey, D. (1993) "Power geometry and a progressive sense of place," in J. Bird, B. Curtis, T. Putnam, G. Robertson, and L. Tickner (eds), *Mapping the Futures: Local Cultures, Global Change*, Routledge: London.

Nordin, U. (1993) "Second homes," in H. Aldskogrus (ed.), *Cultural Life, Recreation and Tourism* (National Atlas of Sweden), Royal Swedish Academy of Science: Stockholm.

Olwig, K. F. (1997) "Cultural sites: sustaining a home in a deterritorialized world," in K. F. Olwig and K. Hastrup (eds), *Siting Culture: The Shifting Anthropological Object*, Routledge: London.

Pearce, P. L. (1982) *The Social Psychology of Tourist Behaviour*, Pergamon: Oxford.

Rojek, C. (1995) *Decentring Leisure: Rethinking Leisure Theory*, Sage Publications: London.

Sack, R. D. (1992) *Place, Modernity, and the Consumer's World*, Johns Hopkins University Press: Baltimore.

Sack, R. D. (1997) *Homo Geographicus*, Johns Hopkins University Press: Baltimore.

Sagoff, M. (1992) "Settling America or the concept of place in environmental ethics," *Journal of Energy, Natural Resources and Environmental Law* 12: 349–418.

Selwyn, T. (1996) "Introduction," in T. Selwyn (ed.), *The Tourist Image: Myths and Myth Making in Tourism*, Wiley: Chichester.

Shurmer-Smith, P. and Hannam, K. (1994) *Worlds of Desire, Realms of Power: A Cultural Geography*, Edward Arnold: London.

Stynes, D. J., Zheng, J., and Stewart, S. I. (1997) *Seasonal Homes and Natural Resources: Patterns of Use and Impact in Michigan*, US Dept of Agriculture, North Central Forest Experiment Station: St Paul, Minn.

Tuan, Y-F. (1977) *Space and Place: The Perspective of Experience*, Minneapolis: University of Minnesota Press.

Urry, J. (1990) *The Tourist Gaze*, Sage Publications: London.

Urry, J. (1996) *Consuming Places*, Routledge: London.

16

LEISURE LOTS AND SUMMER COTTAGES AS PLACES FOR PEOPLE'S OWN CREATIVE WORK

Lena Jarlöv

Introduction

Many urban residents in Sweden have summerhouses in the countryside. This chapter addresses the issue of why a summerhouse can be so important to people. As a young architect I trained to design convenient urban dwellings with high sanitary standards and good service and observed that many people leave these well-designed apartments as soon as they have some days off. They go away to their 'second home' which may be a small simple cottage with considerably lower sanitary standards than their urban dwelling. Some do not even have a toilet and the kitchen is basic, uncomfortable, with a wood stove and without running water.

What are they doing there? One main reason for having a summer cottage is 'to have something to do' (Jarlöv 1982). In this place for leisure and recreation people are working intensively most of the time. They are building, repairing the house, building extensions, sheds for garden tools, greenhouses and endless walls of stones. They are gardening, growing flowers and vegetables, pruning fruit-trees and harvesting the crops, collecting berries and mushrooms in the forest or fishing in the lake or the sea, preparing sailing boats. Besides being a working place, the summerhouse also is a place for close connection to what they feel 'nature' to be. For many people the lifestyle in the summerhouse also means a richer social life with more contacts with the neighbours than in town.

An informing line of enquiry would include an investigation of what it means *not* to be able to get such a complementary place for those who cannot afford it. Is it possible to arrange the urban environment in a way that may compensate the lack of a summerhouse? This discussion is based upon interviews with

70 Swedish households conducted in their homes or summerhouses and also on experiences of working on two development projects in suburban areas of allotment gardens in Gothenburg.

Creativity: a theoretical framework

Two focal concepts concern creativity and creative work, recreation and human needs (Jarlöv 1982). People are essentially productive and active; they conceive, design and accomplish practically. During this process they change and grow. Especially where work is felt to be alienating it is through leisure, the sphere of consumption, that life is felt to be 'owned'. Only there the subject can be her/himself, appropriate the environment in her/his own way – if s/he has the possibilities to do so. Too often the poor remain in a flat owned by someone else and may feel alien also at home (Marx 1965). In this perspective the summerhouse may be the only place where they can be productive and really appropriate the environment without feeling alienated.

Work, as an exchange with nature, can be of the most basic significance for human well-being (Fromm 1941, 1955). To be productive, effective, to achieve something is an existential need. S/he might be crushed by the feeling of powerlessness if the self is experienced as totally passive, as an object. In studies of depression and boredom Fromm argued that the feeling to be doomed to inefficiency, total powerlessness, is extremely painful and almost unbearable. He distinguishes two types of stimuli: simple and activating. A simple stimulus evokes a simple and immediate reaction but does not cause any action. It may give excitement, release and satisfaction to the individual, but if repeated above a certain limit it loses its stimulating effect and will not be registered any more. Further stimulation requires that the intensity of the stimulation increases or that it changes its content; a certain amount of novelty is necessary. An activating stimulus, on the contrary, stimulates the subject to activity. It does not give rise to a simple answer but invites active answers: becoming interested, looking and discovering new features with the stimulus. It brings about a striving. Because of the active response there is always something new. The stimulated subject gives life to the stimuli and changes them through always discovering new features.

In this context the summer house as well as other 'productive' homesteads may activate stimuli that offer possibilities to people to be active and creative in work that is not alien.

> Human life will never be understood unless its highest aspirations are taken into account. Growth, self-actualisation, the striving towards health, the quest for identity and autonomy, the yearning for excellence

(and other ways of praising the striving 'upward') must now be accepted beyond question and perhaps as a universal human tendency.

(Maslow 1954: XII)

Creativity is not reserved for genius or artists. The creative activity comes from within the individual and gives her/him the feeling of realising her/his inherent possibilities.

> The creative process is the emergence in action of a novel relational product, growing out of the uniqueness of the individual on the one hand, and the materials, events, people or circumstances of life on the other.
>
> (Rogers 1961: 139)

Human development is characterised by a dialectical interplay between creativity and social change. Creativity may be described as:

> the faculty which always, in some way, results in transformation of the existing reality, be it natural, material or spiritual, and in introducing new elements given by personal specificity and thus in creating a new reality, a new starting point for new transformations.
>
> (Málek 1969: 165)

Creative work/practice requires both reflection and action. The whole individual is involved. Degrees of innovation are not important, the result is unique or innovative. It can be made according to a model or standard as well as according to one's own ideas and different degrees of negotiation with available cultural models.

Creativity can be material as well as immaterial. The result can be a table-cloth, a building, a dish or a letter to the press. It can also be a solution of a conflict, bringing up of a child or mobilising a group of people to a special action. Every creative activity begins with an unstructured material: a piece of stuff and some yarn to make a tablecloth; a piece of ground and a heap of wood or bricks and other building materials to build a cottage; a piece of butter, some flour, cream, lime and spices to make a sauce; a lot of inarticulate thoughts and words to make a poem; and some angry, discordant and suspicious individuals to form a positive, efficient group of people. Creative work can be summarised as using an unstructured material with input from reflection and deed to reach a structure, a result, with higher or simply different quality than the starting material. The ability to anticipate such a result and to design an activity leading to it, the capability of creating, distinguishes human practice (Jarlöv 1984: 107).

The concept of recreation is often regarded as opposite to work. For many people work *is* recreation, i.e., empowered, free work is recreation. Recreation can be defined as a situation where you are occupied with re-creating your self, with components of material needs (food, housing, clothes), emotional needs (human relations, to be involved etc.), existential needs (contact with dimensions above intellectual understanding, sense of meaning, possibilities to be creative). In our 'consumption culture' people who cannot satisfy their emotional or existential needs may be compensating this with material consumption, or consumption may assist or conceal creativity.

Leisure practice as creative work

A series of explanations of the phenomenon of summerhouses emerges. First, there is a felt need for change. Paradoxically, there are very strong indications that the summerhouse represents the *real* home for many people, where they would stay all the time if possible. Whilst they enjoy travel and visiting places as much as anyone else, their summerhouse represents the firm point in their existence (Jarlöv 1982). What qualities in a place enable the identity of the self in the complex dialogue between self and environment? Having a summerhouse might be a comfortable place for 'laziness' in a particular moral geography. For many people the summerhouse is a place for their own work, 'real work', instead of the alienated labour of employment. In a study of people who had moved from urban housing areas to settle permanently in a summerhouse in the countryside, the possibility for their own work proved to be one of the main motives (Jarlöv 1982). The physical living environment often contains undeveloped possibilities for creative human work. Utilising these possibilities can empower and perhaps offer rest from the hardship of everyday life and time for reflection and meditation.

Many people who live permanently in summerhouses were born in an old central part of Gothenburg or moved in there from the countryside, relatively satisfied with the environment, but were later moved voluntarily or by force to a block of flats in a suburb in the postwar decades. Dissatisfied with their surroundings and social life there they soon started to look for another place to spend at least their holidays. They found a place to build a summerhouse, for some to become their permanent home (Jarlöv 1982).

Contact with nature and the seasons; gardening, building, social life, each of these full of activities; friendships with helpful neighbours; each of these was recalled as missed from previous residences in city and country. Freedom was especially missed and valued: to make what you want, to be yourself, to follow your own ideas, that they felt difficult in modern urban areas. In summerhouses people were extremely active, their homes real working-places.

Lövgärdet is a 1960s suburb surrounded by vast areas of rocky forests not especially appropriate for productive gardening. However, as evident in rocky areas all over the world, people can make the most wonderful gardens if they really want to. Lövgärdet became the site of local activity to develop an allotment garden called Lövköjan. The idea of allotment gardens is old and well established in Sweden (Crouch and Ward 1997). Usually the allotment gardens are available across the city, but here available to those living nearby, walking distance from their flat. This offered opportunity for different family members to have easy access and choice of staying in the garden or in the flat, flexibility often not available to summerhouse owners where their main residence is distant from their flat. Interested inhabitants organised the allotment garden area with the owning housing company close to the housing estate.

The plotholders express many different reasons for having an allotment garden (Bergquist and Jarlöv 1988): to be near to 'nature', 'in peace and quiet'; to be active in work, creating one's own environment, to build and garden; to work with neighbours, to co-operate. The inhabitants of 'Lövköjan' stay in the suburb of Lövgärdet at weekends and most of their holidays instead of going away to another place, unlike those who own summerhouses or have a leisure lot in one of the ordinary allotment garden areas distant from home. This project is felt to strengthen the identity of the suburb and further empowers inhabitants to get together. In spring 1996, the plotholders bought the allotment garden area from the housing company and became owners.

In another project of suburban renewal in Kortedala, Gothenburg, tenants developed an allotment garden (Gunnemark, Jarlöv and Lennartsson 1991) In this area, 'Kajsas ängar' (Kajsa's meadows), many 'unemployed' (sic) people spend much time there, working for themselves. One couple built their 'dream house', like a modern villa in miniature. The husband is 'unemployed' after having been a building worker for many years. They have both been engaged in the group from the very beginning.

The wish for creativity and appropriation of a little piece of land is a very individual desire. The strong social networks characterising most allotment areas don't develop until the gardens are established. Gardening emerges as an expression of identity, one's own place, devotion, tradition, social life, voluntary seclusion, a place with its own rhythm instead of clock-time, the felt importance of work (Bergqvist and Jarlöv 1988, Bergqvist 1989). People out of work because they were considered to be ill, many early retirement-pensioners, work their garden intensely.

Amongst families permanently settled in their summerhouses a number of wives had been persuaded by their husbands to settle. Many wives in this study told about the transitional time with two homes as very laborious, repeatedly packing and unpacking, planning for food and children's clothes, leaving town on Friday afternoon and coming back on Sunday night, never having time for anything else. In that regard the permanent settling in the summerhouse made life easier, but so did the nearer option of an allotment with outhouse. Married women expressed a deference and an irritation to this gendering of practice: 'He must have somewhere to be'; 'He must have something to do'; 'a healthy, strong man has to work with his hands'. Women living alone valued the summerhouse less as 'leisure houses'; they wished to be able to live there permanently (Jarlöv 1996).

Conclusions

Leisure constitutes a means of knowing the self and of access to many dimensions of human life. Spaces in which leisure happens are part of this knowledge that is significantly constituted of practice, but also of desire. The modern urban project has separated work and dwelling. People can adapt to this way of living and they even like it; others wish for another way of living. The solution for some of them has been to make a summerhouse a second home, where they can be active and creative and put their own traces in the environment. Some of them convert the summerhouse into their permanent 'only home'. Enabling small-scale handicraft building and self-supporting gardening, as allotments, can be progressive. As leisure brings self-knowledge and realisation of numerous values, empowerment can be found in reorganising the useful pleasure of one's own work.

References

Bergqvist, M. (1989) *Trädgårdens betydelser för folk i flerfamiljshus* (The meanings of the garden to people living in blocks of flats), Gothenburg: Dept. of Architecture, Chalmers University of Technology.

Bergquist, M. and Jarlöv, L. (1988) *Kolonistuga hemmavid. Utvärdering av Lövkojan – en närkoloni i Göteborg* (Allotment garden cottage near to the dwelling – an evaluation of Lövköjan, an allotment garden area close to a housing estate in Gothenburg), Gothenburg: Dept. of Architecture, Chalmers University of Technology.

Crouch, D. and Ward, C. (1997) *The Allotment: its landscape and culture*, Nottingham: Five Leaves Press.

Fromm, E. (1941) *Escape from Freedom*, New York: Rinehart.

Fromm, E. (1955) *The Sane Society*, New York: Holt, Rinehart & Winston.

Gunnemark, K., Jarlöv, L. and Lennartsson, H. (1991) 'Stadsförnyelse – idé och verklighet. Om ett planeringsexperiment i en 50-talsstadsdel' (Urban renewal – vision and reality. About a planning experiment in a district from the 50s), Stockholm: The Swedish Building Research Council, R4: 1991.

Horgby, C. and Jarlöv, L. (1991) 'Lägenhetsträdgårdar – ett nygammalt sätt att förbättra bostadsmiljön' (Gardens for appartment residents – a new-old way to improve the living environment), Stockholm: The Swedish Building Research Council, T1: 1991.

Jarlöv, L. (1981) Vad betyder olika faktorer i bostadsmiljön för efterfrågan på fritidshus? (What do different elements in the housing environment mean for the demand for summer houses?) Stockholm: Report to the Governmental Committee for Summer Housing.

Jarlöv, L. (1982) Boende och skaparglädje. Människans behov av skapande verksamhet – en försummad dimension i samhällsplaneringen (Housing and joy for creativity. The human need for creative activities – a neglected dimension in the physical planning), Gothenburg: Dept. of Architecture, Chalmers University of Technology.

Jarlöv, L. (1984) 'Kvinnoperspektiv på fritid och närmiljö – skapande verksamhet i boendet' (A female perspective on leisure and housing environment – creative activity in the residential area), in Det nya vardagslivet (The new everyday life), Nordisk Ministerråd in co-operation with Nordisk Råd, Copenhagen, Oslo, Stockholm, Helsinki.

Jarlöv, L. (1991) 'Lägenhetsträdgårdar – en åtrerpptäckt princip att arrangera fler-familjshusens utemiljö' (Tenants' gardens – a rediscovered principle to arrange the environments of the housing), Tidskrift för Arkitekturforskning (Journal of Architectural Research) 4(2): 103–6, Gothenburg.

Jarlöv, L. (1996) 'Stadsekologi och trädgårdskultur. Porträtt av en kretsloppspionjär – Leberecht Migge 1881–1935' (Urban ecology and garden culture. Portrait of a pioneer of recirculation – Leberecht Migge 1881–1935), Stockholm: The Swedish Building Research Council, T1: 1991.

Leberecht, Migge (1981) Gartenkultur des 20. Jahrhunderts, Fachbereich Stadt und Landschaftsplanung der Gesamthochschule Kassel: Worpsweder Verlag.

Málek, I. (1969) 'Creativity and social change', in A. Tiselius and S. Nilsson (eds), The Place of Value in a World of Facts. Proceedings of the 14th Nobelsymposium, Stockholm, September.

Marx, K. (1965) 'Ökonomisch-philosophische Manuskripte', in Sven Eric Liedman (ed.) Människans frigörelse (The Liberation of Man), Stockholm: Aldus.

Maslow, A. (1954, 1970) Motivation and Personality, New York: Harper.

Maslow, A. (1962) Towards a Psychology of Being, Princeton: Van Nostrand.

Prokop, U. ((1976) 1981) Kvinnors livssammanhang. Begränsade strategier och omåttliga önskningar (The life context of women. Restricted strategies and excessive desires), Stockholm: Rabén & Sjögren.

Rogers, C. R. (1961) On Becoming a Person, Boston: Houghton Mifflin.

17

KNOWING, TOURISM AND PRACTICES OF VISION

Mike Crang

Events, knowledge and practice

Tourism is often spoken of in terms of the economic effects it generates, the cultural and environmental impacts it may cause or the system of values it promotes. In this essay I want to suggest that all of these approaches lack a sense of tourism as practice. That is because, with one major exception, they lack an open-ended theory of social events. I want to try and sketch out what such an approach might look like by thinking about the practices of photography, taken from my own work on heritage attractions. The essay focuses on photography for two reasons. First, ideas about images and sight have become paradigmatic in talking about tourist experience. Second, accounts using these ideas seem to rapidly pacify tourists – that is they tend to experience, perceive, receive but not *do*. To rectify this tendency, this essay will suggest we need an open theory of practice. Currently tourist practices tend to be reduced to either bald accounts of structure and values, as though these stand apart from the activities, or to empirically measurable content. In the latter case, it may seem odd to say tourist studies tend to lack a sense of events. Indeed one could say that, in their often empiricist vein, they are totally limited to an event ontology – where all that is admitted as being real or significant are transactions, movements, arrivals, departures, entrance numbers, or overnight stays. However, in all these cases the 'events' are closed replicable actions. They become black boxes into which all the untidiness of life can be quietly stuffed away out of sight. All sense that events might be about change and transition seems lost.

   The exception to all this is a processual form of anthropology based around the work of Victor Turner (1980, 1982). I draw some inspiration from that school, but in itself it has too often served to reduce action to structure – where every event becomes a controlled transformation moving from one predetermined state to another – producing, appealingly for geographers, a symbolic landscape marking particular states and transformations. Thus even the category of the ludic (imaginative time), that most necessarily slippery of states, defined as an interstitial time, becomes contained and constrained. It may be argued that this process of control reflects the system on the ground but it seems a predilection of tourism theories to find order. Perhaps due to a sort of intellectual inferiority complex there is a 'more scientific than thou' feel about some studies. This insecure scientism seems manifest in the wish to create a vision of ordered practices and an orderly practice of academic vision. Let me recount an anecdote, where I attempted to write a piece about reflexive heritage practices for the *Annals of Tourism Research*. I emphasised that the piece was in the first person to show that my reflections on events were just as embedded in the situations as those of other actors; I did not stand over the field I studied, my interpretation was part of the same field of practices (both figuratively and literally). The proofs returned, all neatly copy-edited, to the third person – including the memorable phrase 'for this reason the author has written the piece in the first person'. Increasingly sharp exchanges produced a modified concession (see Crang 1996b). The reason given for the obduracy was that the fledgling discipline sought scientific credibility – even if the editor was sympathetic. We might say the effect is to create a proper place from which to view tourists, but with the side effect that only proper behaviour is visible – in what de Certeau called the 'empire of the evident' (1984: 201). Interpretive practices are set apart from the field they comment upon and stand over it serving to 'purify' both. The result is a loss of the sense of temporality as eventfulness from both theory and the world.

Instead this essay tries to reincorporate a sense of temporality in events. To give a specific focus, the events are photographic ones. In part as I have suggested this is to pick up on what are often remarked upon as key dynamics of tourism as part of an 'image-society'. The argument comprises three main axes of questioning: first the role of photography and images in tourist sites; second, the role of photography in tourist practice; third, the role of photography in assembling a sense of self. Moving along these lines I hope to open up some ideas about tourism as a knowing practice. I choose these words with care since while it is common to see debates of education versus entertainment – or even concepts of infotainment – I want to stress that tourism is about producing knowledge. I want to suggest that even if we accept, and I am not sure that I do, Urry's (1990) division of mass and romantic tourism, then the former still produces knowledge – though not of a kind generally recognised in academic accounts.

Sacralising sites

Following the work of John Urry (1990), one of the more recent fruitful avenues of research into tourist experience has focused on the role of sight, images and pictures. Recent work has problematised 'sight-seeing' and picked up threads of other work, from anthropology on the mythical and from cultural studies on the semiotic economy of tourism, to produce a fascinating mixture. Tourism is seen as an activity embedded in webs of signification. Rather than a narrow motivational psychology, founded on behaviourist ideas of the subject, these approaches have opened up ideas about the entanglement of the social and psychological. Notably this has been the effect of MacCannell's (1976, 1992) reworking of Goffman's ideas about regionalised performance, into back and front stage, to look for a structure of knowledge and desire in tourism. MacCannell takes the tourist as an ideal type for Western society. This blurring of the individual and social allows him to suggest that the modern world offers fragmented and chaotic stimulation. MacCannell argues that it creates a deep seated need for holistic experience – a quest for authenticity. This maps into and is reproduced through the symbolic economy of tourism. I will briefly outline the ways tourism might create 'sacralised' sites, through the semiotics of marking and the practice of photography, to suggest how each has contributed to this vision of tourism.

Spatial semiotics

Tourism can also be seen as a more static spatial semiosis marking out places and objects for special attention, defining the sights to be seen and thus making sights out of sites. This has been linked into accounts of signification deriving from Saussure, the whole assemblage of directed gaze and object forming a sign composed of a signifier and signified. In the case of tourism, the signifier is some form of marker which directs a gaze onto an object: a guide, a brochure, or a plaque would all serve. If tourism is about attaining some knowledge of a world, which is not so fragmented as the contemporary melange, then typical objects would be those located in pre-modern society, those from exotic cultures defined as non-modern – and we have something close to a theory of 'auratic objects' where tourism is about seeing the genuine article. The sense of 'genuine' is thus articulated in opposition to a sense of mass production and replication (Benjamin 1978). The auratic object, and Urry's idea of romantic tourism, fit together rather well with ideas of unmediated and direct experience which draw on an upper class aesthetic (Bourdieu and Darbel 1991). And yet, it is reproductions that define senses of what is genuine, as where Albers and James (1988) sketch out the importance of picture postcards in creating the expected appearances of 'Othered' peoples (see also Edwards 1996; for the role of different kinds of images see Bate 1992, Fryckman and Löfgren 1987:129). The system of signification sketched out thus forms a paradox. On the one hand, it points out what ought to be seen, what is significant. On the other, it intrudes as a sign of mediation, indeed the very presence of other tourists can threaten the object's aura. The auratic object thus forms an unobtainable goal, and tourism can be seen against this vanishing horizon of authenticity (Culler 1981, Frow 1991). In many ways tourism becomes a pilgrimage to mourn that which it destroys. It is also through this action that the expansive spatial economy that marks out its modern incarnation is set up.

Morris (1988) suggests that the intrusion of markers helps explain why the sense is of events no longer being self-sufficient. There can be no self-present auratic experience or object. They are constitutively split between the signified and the signifier. Thus we may set out to see something new but all we see are more markers of tourism. MacCannell (1992) offers a different inflection of this process. He suggests that another displacement of this lack is into seeking the backstage as the authentic and unsigned. Of course, this results not only in the encroachment of tourism into ever more spheres of life, but also the staging of authentic scenes. We can follow a well-worn argument which suggests that putatively spontaneous, back-stage, traditional or whatever events become as carefully scripted and performed with just as careful an eye to the tourist audience as anything else (see Greenwood 1977). The outcome is to allow the slippage of signified into signifier. The appearance or signifiers of authenticity become as important as anything else – and hence to themed heritage experiences and museums without collections (Hewison 1991). However, the critical game of trying to establish 'academic' authenticity is akin to the tourist field rather than its antithesis, being one more attempt to define the auratic object (Crang 1996b).

Photographic economies

It is here that photographic practices often seem to amplify what we could call a vicious hermeneutic circle. They set up expectations or frame what will be seen and the visited sites are measured against the prior expectations. Images of a destination (classically the brochure) set up what is significant. The tourist then departs to actually view what is shown as significant, taking pictures of themselves in front of it. These pictures, or more professional postcards, are then taken or sent home to memorialise the visit. A structure is created, where the pictures circulating around sights are more important than the sights themselves, contributing to 'a kind of alienation which has become a prototypical hallmark of photographic "seeing" in tourism' (Albers and James 1988: 136; cf. Redfoot 1984). A simple anecdote may help here; one of the most famous icons of Vancouver are the 'totem' poles or painted lodge poles which have been on display in Stanley Park since times when exoticising First Nations was regarded as far more acceptable. The poles stand in one rather over-walked corner of a very large park with beaches and woods, yet dominate the postcards. Wandering in Stanley I purchased one such card showing the poles set in a misty background. As I turned to look at them there was a family group arranging

Plate 7 Stanley Park, Vancouver, Canada. The visitors are taking turns being
photographed in front of the poles

Source: Mike Crang

themselves to be pictured standing in front of the poles. (Of course I did what
any reflexive academic would do and took a picture of this process too).

The signs that mark out what is to be looked at become as, or more, important
than the sites themselves. The signifier slips free from the signified and it is the
markers that create the experience, rather than any authentic engagement with
the landscape. I have elsewhere (Crang 1996a) used the vignette of the most
photographed barn in America in Don DeLillo's *White Noise* to illustrate this.
The characters approach 'the Most Photographed Barn in America', stand at the
prescribed viewing point, along with many photographers, and have the option
of buying a postcard taken from the spot. Gradually it becomes clear that the
signs proclaiming that this is the most pictured barn are not matters of report,
but part of the cause; the pictures taken do not record an auratic object so much
as create one. It is not too hard from here to see connections to theories of an
image-based society where all sense of a referent is lost: in Eco's (1987) phrase
a 'hyper-real' world where the pictures and simulations have greater significance
in the system than the event itself.

The arguments over authenticity then come back with renewed force.
This loss of authentic life is linked to the mediation of existence by images. The

spectacle has particular spatial consequences organised through tourism. The world coalesces on a logic of reproducibility (Stallabrass 1996). Sites become marked in terms of their inadequacy to pictures, that then circulate in an inter-textual world. This sort of regime means that flows of visitors ostensibly visiting authentic, original sites are actually 'fundamentally nothing more than the leisure of going to see what has become banal . . . The same modernisation that removed time from the voyage also removed it from the reality of space' (Debord 1967: 168: cf. Robins 1991). A world is created comprising multiple fragments; each site becomes a point in an endless signifying chain of sights, till all it can signify is its incompleteness – an effect Dienst, drawing upon Derrida, called "program-matology" (1994:138). This is a vision where media serve to deterritorialise society – converting places into exchangeable sights. Each sight becomes a frag-ment subject to exchange which means not only are they substitutable but none actually stand as an originary object. Rather than the global village of organic wholeness, the global media produce a sense of the serial distance of each sight from the next. Deleuze (1993) argues that this practico-inert conception of space allows a model of quantitative, replicable instants, which differ only in terms of their location, to become the model for a kind of time marked as infinitely divis-ible and infinitely extendible (Adam 1995, Castoriadis 1991). This kind of time and space lies behind the fragmentation of the world into scenes that offer 'an uninterrupted field of potential pictures' (Galassi 1981:16–24). This means time works not only to create events in time, that is to temporally locate them as part of sequences, but also to frame the events as being of a particular kind of time (Adam 1995).

Indeed, this re-casting of the experiential world might be thought of as a trans-formation of landscape. The estimated sixty billion pictures taken each year would be points of light on the globe, densely clustered round iconic sights and trailing off into a darkened periphery (Stallabrass 1996:13). The bright lights of touristic knowledge would reveal, and obscure, a different version of the landscape. The strength of MacCannell's ideal typical tourist is that it is a figure for a wider state of society. Tourism is perhaps then the ur-text of modern living – connecting to visions of global media and an 'information society'. However, I find this extrapolation rather dissatisfying. It says much that appeals and serves to move beyond simple behavioural models of motivation, and yet it erases the activity and knowledge of tourists. The emphasis on the regionalisation of one semiotic economy – into marked sights, into back stage and front stage and so forth – is surely part of the tale, but only part. These spectacular spaces are linked to accounts of bombardment by images, flows of information and tourists. We have an analytic effect where the actual tourists become rather marginal to the model. Analysts like Deleuze or Debord have in effect 'put phenomenology in reverse, spewing the inward outward, forcing consciousness to become a wandering orphan among the things called images' (Dienst 1994:148). We have to try and distinguish between the de-territorialising effect of reducing sights to photo-graphic symbols and the de-territorialising effect of a certain interpretive optic.

Even descending from the grand sweep of high theory, the practices of tourism still tend to disappear in accounts which prioritise the images produced rather than the practices producing them. This is the typical strategy of a content analysis of tourist brochures or postcard or, more rarely, actual tourist pictures themselves (Dann 1996; Löfgren 1985, 1985). Even iconographic accounts drawing on the grand traditions of anthropological structuralism, interpreting visual and material culture (Ball and Smith 1992) also stop at this point. Stopping at the frozen image, leaves us with at best the trace of practice that occurred or occasionally then with some a study drawing on comparative mythology and iconography. It was this conundrum that faced Michael Lesy in his study of family pictures:

> You begin to lose track of specific faces and places, and you begin to think about genotypes. In between the howls and the whispers and the cosmic laughter, you catch yourself short and start thinking about pictures as symbols; and once you start on symbols, you get into a lot of trouble with most civilians. . . . Because it is not long before you start reading comparative mythology. Which does about as much good as the study of chivalry did Don Quixote.
>
> (1980: xii)

Rather than thus seeing pictures as symbols of some hidden meaning, we can think about the practices involved in producing them. Restoring some sense of eventfulness to the pictures can then be achieved without being seduced by the amnesiac geographies of flows, where the experience of the actual tourist seems all too easily erased. For instance, Virilio's (1989) account of the geomet-rification of vision through media of differing sorts thus tends to reduce places to image sites. To use de Certeau's (1984) terms, current models all too often offer places for actions that are somehow drained of temporality, and flows that come adrift from the events that produce them. Instead I want to reinscribe tourism as a space of activity, as a practice producing knowledge not necessarily in the form registered by a theoretical gaze.

Practices of observation

A richer account of viewing practices comes from the work of Krauss (1988) and Crary (1988, 1990) where the focus is on how differing technologies of viewing, comprising the material instruments, their ideologies and the comport-ment and disposition of viewers, create economies of truth and knowledge which go on to become distinctive, rather than unified, forms of visual consumption. The changes in modes of representing the world become part and parcel of changing social formations and practices (Green 1995). At its simplest level, the practices of observation reveal the social dynamics of tourism. In a straight-forward manner we might explore the varying permutations of observation in the tourist situation and its mediation by the presence of a camera. Cohen, Nir and Almagor (1992) examine in detail the possible interactions of strangers and locals in photographic practice, highlighting the way any idea of a singular gaze needs considerable qualification. While assumptions tend to be that pictures are taken of unwitting locals or oppositely of locals performing especially for the tourists, we should also remember the possibilities of guides or 'tourees' shaping pictures. More importantly, we should recognise that the tourists are very often the objects of their own gaze – where tourists depict 'life on the road' or they or locals take pictures of their act of visiting. This is not necessarily celebrating the spectacle of tourism, with the ironic social reflexivitiy of the post-tourist, but also the common picture of tourists in front of a well known symbol of the area visited. Thus Bourdieu (1990) notes how the scene can dominate the visitors, since fitting in the scene is more important in composing the picture than clarity about the visitors. Although the tourists are the objects, the scenery functions to authorise the visit, and the activity depicted is not one of gazing on the landscape but turning backs to it and returning the gaze of the camera (Redfoot 1984). The photographic subjects of a travel album may well not be the album's theme, rather their significance to the album maker (who may or may not be the photographer) is the real subject of album (Walker and Moulton 1989: 173). It can be useful and informative then to think through the multiple permutations of who is picturing, what or whom, where and with what context of later exhibition, circulation or display, either in mind or happening as the researcher comes across the pictures (Crang and Cook 1995; Chalfen 1987). Whereas typologies of interaction scenarios and pictures boil the event of visiting to its bones, I want to use them as an entry point to open up the times and spaces of practices. We have a sense of audience and performance here. The visitors are using the picture to say something, as a communicative tool, to reach people distant in space and time. Tourist photography is a practice directed towards these people and places as much as towards the landscape in front of it. The landscape depicted is thus, as noted above, one marked and composed for tourism. It is a process that requires a more attentive study of practice.

If we think of the role of markers and signs it becomes clear that the land-scape to be observed is shaped in particular ways – something carved out as an object – where sights are located and em-placed by a particular gaze. There is a connection between the scenic postcard or picture and ways of seeing in terms of detachment and objectification of the landscape (Green 1995: 37–8; Löfgren

1985: 92). Touristic pictures promise a world where the 'co-ordinates of cultural identity and comparison are fixed in small easy-to-carry squares. It is as if a great reduction machine were at work turning life into a billion minatures' (Hutnyk 1996: 147). The creation of this landscape through ways of seeing can be conveyed through Heidegger's terms of enframing or enworlding. Heidegger's schema moves us forwards since neither knowing nor known are independent terms, but instead the form of knowledge is linked to the practices through which it is created. In this sense he alerts us to a *pre-ontology* of what is made visible or apparent by a particular practical stance offering some purchase on the predisposition of tourism, as outlined above, as well as the process itself. That is, objects are offered up to be known in particular ways linked to how they are approached. The work on the general semiotic economy of tourism emphasises the cutting out of places from their original context by the markers of tourism, their reframing in the specifically created contexts of tourism. This commodification and presentation in Heidegger's terms is making places 'occurrent'. This has a particular significance since occurrent indicates something not merely occurring but set out as a specific thing to be experienced – it is in his terms 'inauthentic'. It is related to a way of being-in-the-world that carves out events from their embedded context. The idea of occurrence thus suggests events do not simply happen but are set up to happen.

Creating spaces through practice

The idea of an occurrent landscape suggests seeing the creation of sites as a creation of events waiting to happen. Certainly we need in some way to reflect that, despite being a practice with very little formal coaching, there are few cultural products more standardised and less reflective of the anarchy of individual intent than photographs (Bourdieu 1990). Some have used this as a reason to look for a basic psychological grammar (Hirsch 1981; Ziller 1990). Instead Bourdieu finds a logic in the practice that avoids a retreat into ahistoric essences. As he points out, the manuals of tourism and photography coalesce into a call for a sort of constant curiosity that differs markedly from everyday inattentiveness. This discourse does not stand apart from the practices but enters into them – in the form of not just guides about what to look at, but scripts of how to do so. Thus photographic companies have spent almost a century telling the public how to compose pictures and assess their function. This is a social scripting working through the way we have learnt to use a technology, the way sights are set up and how we shape events to provoke its use ('let Kodak do it for you'). It is thus a matter of historical evolution and change where the logic of the photographic industry is also involved. As Crary (1990) has argued, there is no one simple master story about ways of seeing but rather a densely patterned interlocking web of technologies of vision and practices of observation that form a complex historical-geography of sight. The photographic industry makes most money from film rather than hardware. The effect of which is to make it more profitable to get average users to use average cameras a little more rather than promote aesthetic or technical innovation. The interests of the industry have produced an advertising and marketing direction that has consistently stressed features that make a camera easy to use. Kodak went global with the slogan 'Point. Shoot. We do the rest' almost a century ago. Nowadays most advertising still highlights features such as auto-focus. The century has not been about campaigns of technical training or skills but simplicity and instantaneity of use marketed through the idea of self-expression.

It has also served to colonise more and more experiences to produce a growing market under the rhetoric of spontaneous snapshots. From formal portrait photographs to holiday snaps, from domestic rites of passage to family events, the amount of life enframed by photographic vision has increased, and while less and less is formally posed we might wonder whether more and more is not always ready to be snapped. There is what Tagg (1990) calls the dialectic of self-expression and colonisation of the lifeworld. However, in restating this trend we risk repeating a quest for authenticity – where we nostalgically appeal to some authentic unmediated type of experience as a contrast. We risk moving from seeing representations and images overwhelming sites to seeing a form of photographic practice corrupting some original model of experience. I want to suggest instead going with the grain of practices by looking at tourist photography as a knowledge producing practice, undeniably situated in specific ways but which needs understanding rather than denunciation. In this sense I want to develop a notion of reflexivity within the practice; it is a self-knowing operation in the sense of constructing a story of the self, and between it and academic understandings where we might look to the parallels of two knowledge producing systems. Trinh Minh-ha (1989) has argued that we might usefully see anthropology as 'gossip', in terms of a profuse idle chatter about someone – a production of knowledge through chatter (cf. Crick 1992). Likewise we have to note the chattering and storied nature of tourism, both in meetings of travellers and in later recountings of events. The idiom and props for this sort of knowledge production are often photographic (Hutnyk 1996: 62–4, 145; Murray 1993). Perhaps then, we should look at pictures as opening up narratives around the moment captured (Price 1994: 40; Walker and Moulton 1989: 168). I want to argue that relationships of time and space are configured in a particular manner through this knowledge production system.

Time and space

The creation of an occurrent landscape though is only one element of the times and spaces of tourism. It is the creation of so many places, whereas after de Certeau (1984), we might see space as practised place. Analysing inscription and marking, without looking for practices, can only produce a mortuary geography drained of the actual life that inhabits these places. Interestingly though this sense of time lost seems as though it could be used productively to tell us about tourism. Photographs are often argued to be a necessarily elegaic object – tearing a moment of the past out and preserving it into the future. We might see obvious parallels with the sort of temporal manip-ulation engaged in by museums or recreated tourist experiences (Crang 1996b) and indeed, with the fortuitous moment when a visitor shows a picture of York in fog, which manages to crop out most 'modern intrusions'. But the ways in which photography connects personal time to tourism seems a more rewarding avenue to pursue here. Tourist photography replicates the semiolo-gical economy outlined above, in the way it denies any given moment's self-presence. Tourism is often criticised for the way it generates inauthentic, other-directed behaviour among locals – where the event or culture stops being-for-itself and is directed for outsiders. Photography also enters into events so they are not for-themselves. Pictures do not simply record events; the practices of photography are part of them. The events become mortgaged to the future, where self-presentation is not simply about current companions or audiences but future ones removed in space and time. Photography provides a tech-nology through which space and time bound events can be made spatially and temporally portable.

This spatio-temporal displacement needs to be followed through carefully. The times and spaces involved can be varied and complex in both empirical and theo-retical senses. If we sketch out some of the lineaments of this topology we can see some key trends. We need to think of the intended audience. These may be the participants themselves – in terms of using mementoes and souvenirs to bring back memories. The picturing acts to make the present appear as a fore-shadowing of its own reminiscence. This is very much the message of successive advertising campaigns by companies like Kodak, with slogans like 'Don't let these precious memories slip away' and visuals showing depicted matter fading to grey, evoking the transitoriness of experience against the relative permanence of photographs. Human memory and actual experience is portrayed as fallible and inferior to its celluloid trace. My favourite evocation of how taken for granted this devaluation of unaided memory can be is the shock I felt when reading the following anecdote:

> When Joe Louis was asked shortly before he died whether he would like to see a film of his greatest triumph (the second fight with Max Schnelling), he tersely declined: 'I was there.'
>
> (Murray 1993: ix)

It is not only a matter of personal memory, the audience can also be other people, be they relatives or friends (present or future). In which case we need to bring in dynamics of status display; the easiest (clichéd) example would be showing off holiday slides to neighbours. While there are clearly differences between the visitor who described trawling through remote Polish railway lines, and taking illegal pictures from the car in the former East Germany, to catch them 'before they were modernised' and a visitor simply catching a moment in their children's growth, one focusing on the transient object, the other on the transience of life, the future perfect (will have been) sense of photography stretches across both. We might then see this taking the logic of reminiscence further until being there is less crucial than having been (Kelly 1987, 1986). Of course, a similar function for a spatially removed audience can be performed by postcards – where the message written may be 'wish you were here' but it is also clearly, 'I was and you are not'. Snapshots are also temporally mediated forms of display, in terms of the social trajectory of the takers. Thus they may be used years or decades after the event when the social circumstances of the keepers may have changed considerably.

This focus on self-image manipulation seems too instrumental a vision of tourism. Self-awareness and self-presentation should not necessarily imply a nexus of social calculation and cynicism. The ideology of photographs is also one of authenticity – preserving the instant, unposed, and spontaneous. This works to make people already aware of the possibility of taking or being in pictures and comes to be part of normal self-presentation. We might look to a middle-class 'Kodak culture' (Chalfen 1987) where photographs are used to affirm the values of success, earned leisure and familial happiness that are privileged by that class fraction. Similarly Bourdieu (1990) discusses family photography where the picture both celebrates the object (the family) and the practices serve to reinforce the rituals of the family. So we might find a similar circularity, without intentionality, in touristic pictures. As Sontag put it, '[i]t seems positively unnatural to travel for pleasure without taking a camera along. Photographs will offer indisputable evidence that the trip was made, the program was carried out, that fun was had' (1977: 9). Photographs are social evidence of achievement, but pleasurable evidence. While thus far I have stressed the evidentiary qualities of pictures as 'record of achievement' in fact we should be equally alert to the humorous sides. I say this because so often research methods work to stress the functional and instrumental aspects of tourism. Reading study after study of tourism, it is very hard to get a sense of people actually enjoying themselves or having fun. We might call it the social reproduction of seriousness in academic studies.

This account of the after-life or circulation of snapshots does not exhaust the time-space shifts in pictures. If we think of the earlier comments on a hermeneutic circle then we have to see photographic images pre-figuring the experience as well. Not only that but our preparedness to take pictures is a predisposition structuring the experience. A predisposition also mediated by the particular technology that has social implications too. Thus we might ask who is taking the pictures. Certainly some technologies are more strongly gendered as masculine than others – thinking of, say, camcorders. The scene that most brought this home to me was where a family with whom I was working visited a castle with the camcorder I had lent them. The father and son went off filming a boy's own adventure of the son exploring the ruined grounds. As they emerge from the moat one scene was of the mother still sitting on a tumbled wall engrossed in her book. In a later interview the father narrated this incident with the comment that ruined castles to her were a pile of rock, while to the son the opportunities this provided for climbing and adventure were the very point of the sites. There can also be whole hierarchies of taste over cameras – from throwaway one offs, to automatics and professional ones with complex lenses and so forth. The logic need not simply be one of increasing price and complexity; it may be that having too expensive a camera may suggest the photographer is trying too hard, or indeed some may reject a camera altogether in a reflexive gesture aimed to recover authentic experiences (Redfoot 1984; Bourdieu 1990). Meanwhile studies of the use of photographs suggest that women are often more active in their collecting, editing and display (Lesy 1980). And of course the spaces of display and archive are entangled in this discussion. Those pictures that will move into albums, will be different from those condemned to the loft, those on the bedside table differ from the best room's mantelpiece; those on video differ from those made into slides; family slides differ from those that become adapted to illustrate lectures or books. The practices within which these pictures sit are not solely visual, but rather are endlessly storied. I have never been shown a picture in total silence. Indeed, the presence of the 'owner' or 'keeper' with the person looking at tourist shots, as part of an occasion for social interaction, means pictures are offered with preferred interpretations and guidance. Within the heritage business there is a whole circulation of archive pictures appearing in books, in exhibits, in 'ye olde Aletaster' and even in supermarkets.

Pictures allow the possibility of a capitalisation of experience. To use Benjamin's phrase they convert the 'erlebnis' of actual experience into 'ehrfahrung' of

250

deceased experience (Benjamin 1973; Sontag 1977: 7). The practices of photography insert a new form of time into the events. It is not merely that the event becomes located in a series but that seriality enters into the event itself. It is not merely the presented past that is a foreign country but also the event of going to see its monuments – pictures double encode the sense of loss and presence. The possibility of the photograph as a souvenir converts a journey into an excursion (Stewart 1984) where experience is allowed to be accumulated rather than lost, or, rather the acknowledged loss of the experience is what produces aesthetic charge in the picture – perhaps explaining why other people's holiday snaps can be so tedious. This is encapsulated by what Agamben (1993: 33) calls the fetish-paradox: 'the fetish confronts us with the paradox of an unattainable object that satisfies a human need precisely through its being unobtainable'. The context of display also plays a mediating role in the relationship to these lost moments. Very few pictures are viewed singly but more often as part of an album or, even more grandly, an archive, and we need to think how the assemblage of images affects them. Events can be piled up or brought together through a narrative in the sequence of images. This is perhaps most obviously evident in the genre of photographic journeys spawned by the early spread of photography. We can find systematic examples in the practices of imperialist propaganda but equally in early tourism (Bate 1992; Ryan 1994). Thus the photographer August Léon was employed by the financier Albert Kahn as part of his project to create a photographic 'planet archive'. Léon travelled through Sweden and round Lake Siljian in 1910, and his itinerary and experiences were accumulated to form a picture of a particular regional culture as suggested by the exhibition of these pictures, a version of which was published as 'Med hyrbil och kamera till Dalarna 1910' ('Through Dalarna in 1910 with Hired Car and Camera'; Leksands Kulturnämd 1994). We need then to think about the ways structures of journeys and particular technologies of travel intersect with those of representation (Löfgren 1985: 95–6) – where travellers will liken photography to looking from a train window (Hutnyk 1996: 151). A spatial practice of travel built up a picture of a folk culture and a region before the camera lens. Chance events and encounters are frozen as representatives of that culture. Of course these pictures were prefigured by an artistic movement that celebrated the region as worthy of depiction and the pictures (and similar ones) went on to shape expectations of later tours. The pictures thus traced back examples of local culture promoted through Swedish pavilions at world fairs, themselves another version of the world as exhibition, and also recorded the locals as tourees already installed as theatrical guides in museums, and the invented 'folk' costumes of hotel staff, as well as authentic folk costumes then being formalised, redefined and canonised as part of a nationalist movement. This journey brings us back to earlier suggestions of how voyages also set up incidents in an inter-textual system related to each other rather than being self-present. If we turn from the ever receding originary moment and think instead of the practice of an accumulatory project, it opens up possibilities for thinking through the interaction of tourism practices, social formations and self-definition.

Modern stories and knowledge in tourism

The sense of an accumulatory project is not a universal or necessary part of photography, but it is strongly linked to tourism. We need only think of the colloquial phrases of 'doing Italy' or 'doing Scandinavia' to see an energetic inquisitiveness in the process. It is one Bourdieu (1990) linked strongly to his concept of a *new petite bourgeoisie*. For this class fraction, Bourdieu saw a strong ethos of self-improvement typified by acquiring cultural knowledge and symbols. In contrast he suggested a more *haut bourgeois* stance was one where unmediated experience was more highly valued, and was enabled by taken-for-granted cultural competences. Without these competences tourists create a middle-brow form of knowledge that aspires to the same aesthetic criteria but works through self-education. The ability of photography to capitalise on experiences provides an opportunity for the accumulation of cultural symbols which appears to fit this modus vivendi. We can echo Strathern's (1992: x) question as to whether 'this is the last of a long line of middle-class projects, one wonders: middle-class because this is the class that makes a project out of life, that makes *experience out of interactions*' (emphasis added). Given this is the class background of most visitors to heritage attractions the use of cameras seems to offer a way of keying into a mode of touristic experience and knowledge production.

The capitalisation of experience allowed also keys into debates over the relationship of commodification and everyday life. It also offers the possibility that photography is a form of self-narration suitable for the modern age. It is a form of self-creation that is based around a fractured and presentational existence, worked through technologies of representation (Walker and Moulton 1989; Reme 1993: 36). Whereas classical autobiography has been dominated by linear, chronological forms these popular productions might be seen more as spatial stories. Conscious tellings of lives through particular spaces. Seeing the pictures as thus embedded in time-spaces of self-(re-)presentation and narration seems to offer some mileage beyond the world of circulating images so often suggested. We also need to think then of pictures not simply as about loss but also connectivity and contact. The snapshot, like all souvenirs, is not simply a pictorial form but an object. An object that connects us to other times and spaces by its material presence. The logic is not purely metaphoric and iconic but also metonymic – its presence reminds us of a larger whole. So in thinking about pictures we need to see them as props, perhaps as topoi, on which certain memories and events are hung (Csikszentmihalyi and Rochberg-Halton 1981). Or, less consciously perhaps, see them as reminders or madeleines that may trigger unbidden memories of past occurrences. In this way we can use pictures to reinsert an eventfulness and temporality often strikingly absent in totalising visions of tourism.

The academic gaze often replicates the practices of tourism. So much of what is produced is couched in a rhetoric of 'peering beneath the veil', seeing what is really going on – as though that did not also comprise part of the practice being studied. We do not have to accept a total isomorphism of the manner of study and the object of study to see strikingly similar logics. The trip into the field to acquire knowledge, the gathering of material, the souvenirs as strips of deceased experience (Hastrup 1992a and 1992b), the presentation of pictures as proof of having been to authorise our stories (Lévi-Strauss 1973: 17–8), the perpetual present of the moments we talk about and the conversion of all this to academic capital (de Certeau 1988); the parallels are striking. Indeed so are the framing of an academic gaze and the way it too shapes places into sites of knowledge. I hope this essay has set in motion some thoughts about the relationships between forms of knowledge. The phrase 'knowing tourism', is meant to open a threefold field. First, the sense of academics knowing about tourism. Second, the focus on practices also shows tourists as self-knowing. I do not mean this in the sense of a post-tourist with an ironically mocking self-awareness – for which I think the evidence is mixed. I mean it in the general sense of self-presentation and self-monitoring, thinking about our performances and who may see them, when and where – which does not exclude ironic detachment as one among other strategies. Third, I have been trying to suggest that tourism produces knowledge about ourselves and the world. Not immaculate or academic knowledge but knowledge none the less. In order to see this knowledge as active I have suggested we need to look at the practices of tourism and the way they are embedded in and, in turn, transform times and spaces.

References

Adam, B. (1995) *Timewatch,* Polity Press: Cambridge.

Agamben, G. (1993) *Stanzas: World and Phantasm in Western Culture,* University of Minnesota Press: Minneapolis.

Albers, P. and James, W. (1988) Travel Photography: a methodological approach, *Annals of Tourism Research* 15: 134–58.

Ball, M. and Smith, G. (1992) *Analyzing Visual Data,* Sage: Beverly Hills.

Bate, D. (1992) The occidental tourist: photography and colonizing vision, *AfterImage* 20 (1): 11–3.

Benjamin, W. (1973) *Illuminations* (trans. H. Zohn), Fontana: London.

Benjamin, W. (1978) *One Way Street and other writings,* Verso: London.

Bourdieu, P. (1990) *Photography: A Middlebrow Art,* Polity Press: London.

Bourdieu, P. and Darbel, A. (1991) *For the Love of Art: European Art Museums and their Public*, Polity Press, Cambridge. First published in French by Editions Minuit, 1969.

Castoriadis, C. (1991) Time and Creation, in Bender, J. and Wellberg, D. *Chronotypes: the construction of time*, California: Stanford University Press, pp. 38–64.

Certeau, M. de (1984) *The Practice of Everyday Life* (trans. S. Rendall), California University Press: Berkeley.

Certeau, M. de (1988) *The Writing of History* (trans. T. Conley), Columbia University Press: New York.

Chalfen, R. (1987) *Snapshot Versions of Life*, Popular Press: Bowling Green, Ohio.

Cohen, E., Nir, Y. and Almagor, U. (1992) Stranger–local interaction in photography, *Annals of Tourism Research* 19 (2): 213–33.

Crang, M. (1996a) Envisioning the city: Bristol as palimpsest, postcard and snapshot, *Environment and Planning* A: 28 (3): 429–52.

Crang, M. (1996b) Living history: magic kingdoms or a quixotic quest for authenticity? *Annals of Tourism Research*, 23 (2): 415–31.

Crang, M. and Cook, I. (1995) *Doing Ethnographies*, CATMOG 58.

Crary, J. (1988) Modernizing vision, in Foster, H. (ed.), *Vision and Visuality*, Bay Press: Washington, pp. 29–50.

Crary, J. (1990) *Techniques of the Observer*, MIT Press: Massachusetts.

Crick, M. (1992) Ali and me: an essay in street corner anthropology, in Okely, J. and Callaway, H. (eds), *Anthropology and Autobiography*, Routledge: London, pp. 175–92.

Csikszentmihalyi, M. and Rochberg-Halton, E. (1981) *The Meaning of Things*, Cambridge University Press: Cambridge.

Culler, J. (1981) Semiotics of tourism, *American Journal of Semiotics* 1: 127–40.

Dann, G. (1996) The people of tourist brochures, in Selwyn, T. (ed.), *The Tourist Image: myths and myth making in modern tourism*, Chichester: Wiley, pp. 61–82.

Debord, G. (1967) *The Society of the Spectacle*, Black and Red Books: Detroit.

Deleuze, G. (1991) *Bergsonism*, Zone Books: New York.

Deleuze, G. and Guattari, F. (1987) *A Thousand Plateaux*, University of Minnesota Press: Minneapolis.

DeLillo, D. (1986) *White Noise*, Picador: London.

Dienst, R. (1994) *Still Life in Reel Time*, Duke University Press: Durham.

Eco, U. (1987) *Travels in Hyper-reality*, Picador: London.

Edwards, E. (1996) Postcards – greetings from another world, in Selwyn, T. (ed.), *The Tourist Image: myths and myth-making in tourism*, John Wiley: Chichester, pp. 197–222.

Frow, J. (1991) Tourism and the semiotics of nostalgia, *October* 57: 123–51.

Fryckman, J. and Löfgren, O. (1987) *The Culture Builders: an historical anthropology of middle class life*, Rutgers University Press: New Brunswick.

Galassi, P. (1981) *Before Photography: painting and the invention of photography*, New York: Museum of Modern Art.

Gordon, B. (1986) The souvenir: messenger of the extraordinary, *Journal of Popular Culture* 20 (3): 135–46.

Green, N. (1995) Looking at the landscape: class formation and the visual, in Hirsch, E. and O'Hanlon, M. (eds), *The Anthropology of Landscape: perspectives on place and space*, Clarendon Press: Oxford, pp. 31–42.

Greenwood, D. (1977) Culture by the pound: an anthropological perspective on tourism as cultural commoditization, in Smith, V. (ed.), *Hosts and Guests: the anthropology of tourism*, University of Pennsylvania Press: Philadelphia, pp. 129–38.

Hastrup, K. (1992a) Anthropological visions: some notes on textual and visual authority, in Crawford, P. and Turton, D. (eds), *Film as Ethnography*, Manchester University Press: Manchester, pp. 9–22.

Hastrup, K. (1992b) Writing ethnography: state of the art, in Okely, J. and Callaway, H. (eds), *Autobiography and Anthropology*, Routledge: London, pp. 116–33.

Hewison, R. (1991) The heritage industry revisited, *Museums Journal* Apr.: 23–6.

Hirsch, J. (1981) *Family Photographs: content, meaning, affect*, Oxford University Press: Oxford.

Hutnyk, J. (1996) *The Rumour of Calcutta: tourism, charity and the poverty of representation*, Zone Books: London.

Kelly, R. (1987) Museums as status symbols II: attaining the status of having been, in R. Belk (ed.), *Advances in Non-Profit Marketing* (vol. 2), JAI Press: Greenwich, Conn./London, pp. 1–38.

Kelly, R. (1986) International tourism: pilgrimage in the technological age, in Chin, T., Lazer, W. and Kirpilani, V. (eds), *Proceedings of the American Marketing Association's International Marketing Conference*, Singapore, Jun. 16–18, pp. 278–85.

Krauss, R. (1988) The im/pulse to see, in Foster, H. (ed.), *Vision and Visuality*, Bay Press: Washington, pp. 51–78.

Kruse, H., Jobs-Björklöf, K. and Andersson, R. (1994) *Med hyrbil och kamera till Dalarna 1910*, Leksands Kulturnämd: Dalarna.

Lesy, M. (1980) *Time Frames: the meaning of family pictures*, Pantheon Books: New York.

Lévi-Strauss, C. (1973) *Tristes tropiques*, Jonathan Cape: London.

Löfgren, O. (1985) Wish you were here: holiday images and picture postcards, *Ethnologia Scandinavica*, 90–107.

MacCannell, D. (1976) *The Tourist: a new theory of the leisure class*, Shocken: New York.

MacCannell, D. (1992) *The Tourist Papers*, Routledge: London.

Morris, M. (1988a) At Henry Parkes Motel, *Cultural Studies* 2 (1): 1–16.

Murray, T. (1993) *Like a Film: ideological fantasy on screen, camera and canvas*, Routledge: London.

Price, M. (1994) *The Photograph: a strange confined space*, Stanford University Press: California.

Redfoot, D. (1984) Touristic authenticity, touristic angst and modern reality, *Qualitative Sociology* 7 (4): 291–309.

Reinkvam, O. (1993) Reframing the family photograph, *Journal of Popular Culture* Spring: 39–67.

Reme, E. (1993) Every picture tells a story: wall decoration as expressions of individuality, family unit and cultural belonging, *Journal of Popular Culture* Spring: 19–38.

Robins, K. (1991) Prisoners of the city: whatever could a postmodern city be? *New Formations* 15: 1–22.

Ryan, J. (1994) Visualising imperial geography, *Ecumene* 1: 157–76.

Sontag, S. (1977) *On Photography*, Penguin: London.

Stallabrass, J. (1996) *Gargantua: manufactured mass culture*, Verso: London.

Stewart, S. (1984) *On Longing: narratives of the miniature, the gigantic, the souvenir and the collection*, Johns Hopkins University Press: Baltimore.

Strathern, M. (1992) Foreword: the mirror of technology, in Silverstone, R. and Hirsch E. (eds), *Consuming Technologies: Media and Information in Domestic Spaces*, Routledge: London.

Tagg, J. (1990) *The Burden of Representation: essays on photographies and histories*, Macmillan: Basingstoke.

Trinh, Minh-ha (1989) *Woman, Native, Other*, Indiana University Press: Bloomington.

Turner, V. (1980) Social dramas and stories about them, *Critical Inquiry* 7 (1): 141–68.

Turner, V. (1982) Dramatic ritual/ritual drama: performative and reflexive anthropology, in Ruby, J. (ed.), *A Crack in the Mirror: reflexive perspectives in anthropology*, University of Pennsylvania Press: Philadelphia, pp. 83–99.

Urry, J. (1990) *The Tourist Gaze*, Sage: London.

Virilio, P. (1989) *War and Cinema: the logistics of perception*, Verso: London.

Walker, A. and Moulton, R. (1989) Photo albums: images of time and reflections of self, *Qualitative Sociology* 12 (2): 155–82.

Ziller, R. (1990) *Photographing the Self: methods for observing personal orientations*, Sage: Newbury Park/London.

18

THE INTIMACY AND
EXPANSION OF SPACE

David Crouch

Introduction

The purpose of this chapter is to explore the way that spatial practice works. Leisure is one way in which people practice space; Crang has interpreted work-space as another (1994). The way in which this happens in leisure may constitute particular valued components that may make leisure worthwhile, or make it an enjoyment.

Leisure spatial practice

The increased relative commodification of leisure sites has probably had an influence in recontextualising leisure practice. This is most evident in the commercial efforts to include leisure in 'shopping experience'. However, the degree to which this has reformulated what shopping means, and indeed what spatial practice in shopping means, is the subject of critical debate (Miller *et al.* 1998). For over a decade the semiotic power of commercial sites became especially prominent in arguments concerning what happens in leisure. The semiotic power of particular sites (too easily imagined as 'sights') drew emphasis from ideas of the gaze that linked powerfully with earlier geographical emphasis on 'fields of vision'. Of course this position was confronted, especially by feminist critique for its over-sight of other qualities of engagement, especially of embodiment, of the human subject (Rose 1992; Wearing and Wearing 1996). In an early contribution to the critique of the power of commercial context to refigure the content of practice Warren pointed out the importance conceptually of making sense of commercial leisure landscapes and how people make sense of them in practice (1993). Lay geography becomes very important here in terms of the competing ground for different interpretations of practice, and how the human subject 'makes sense'.

Warren's argument worked from Fiske's and Gramsci's claims that people use different agendas from those supplied by commodification and work their own sense into these leisure spaces, also explored by Crouch (1992). This has been considered as over optimistic but at least it serves to unsettle claims that work the other way and place over-arching power on particular messages produced by highly invested symbolism. The contemporary importance of lifestyle suggests that practice works more messily and that there is negotiation and tactic in semiotic consumption (Miller 1987, Chaney 1996), that commercially figured sites, as other goods, may be felt to provide useful metaphors through which to adjust, hold onto, or discover aspects of lifestylisation.

Indeed the investigation of car-boot fairs questions the limits of consumption further. Lay knowledge figured in these spaces and in these practices revolves around a tactical knowledge that includes networking sites, friendships, performance and amusement (Gregson and Crewe 1997). Anticipating more recent work on tactical and socialised knowledge in shopping Ley and Olds observed that ideas of family can focus leisure practice even in highly metaphorical commercialised contexts (1988). The geographical knowledge figured in visiting a World Fair was contextualised in intentions of identity and was practised through familial engagement and play during the visit, where spectacles provided interesting material for this enactment of friendship to use.

There are other metaphors that come into play in terms of leisure practices. These, such as 'nature', city and country, are adjusted by commercial consumption-oriented site production but draw upon other metaphorical sources (MacNaghten and Urry 1998) and may also be refigured in practice (Dickens 1996). Metaphors like these seem to relate especially to contemporary commercial contexts of leisure, even though they draw, unsurprisingly, on modern and even pre-modern notions to do so (Crouch 1992). It would seem especially useful in understanding how people make use of these complex metaphors in notions and practices that surround what they do in leisure, and how these may inform their spatial practice. An increasingly fruitful line of enquiry here seems to be through embodiment in the way space is practised. At least, it would seem that leisure space is not practised merely semiotically (Crouch 1998).

In working through the perplexity of metaphor in leisure spatial practice Ann Game's work on Deleuze, Guattari and Cixous provides important new insights for making sense of lay geography. Reflecting on her own practice in 'doing' Bondi beach and the Yorkshire Dales, Game provides a very felt sense of practising space 'with both feet', 'a desire to know what cannot be seen' (1991: 184). This practice emphatically engages the embodied subject practising space and making sense of where she is, trying to connect with particular available stories and doing so imaginatively. In this way she feels the subject making knowledge. Game believes that although he was criticised for romantic distortion of the possible, de Certeau enables enlarged thinking about proper codes and 'proper' knowledge that has surrounded too much interpretation of contemporary leisure practice, dominated as it had been by a semiotic, consumptionist perspective (as Miller *et al.* argue of studies of shopping (1998)). It is not necessary to engage the limits of guerilla tactics in grasping the potential of the embodied subject to 'make sense', indeed to grasp the world about her. This may require calling upon a complexity of metaphor as well as feeling the ground.

Although she has not engaged in a discourse on embodiment, Finnegan's work on especially leisure-time music activities has the potential to enlarge our consideration to an embodied inter-subjectivity in leisure spatial practice. Finnegan argues for the importance of different processes for leisure from those usually included in 'consumption'. In particular these are ritual enactments that bear a shared meaning amongst participants, and artistic production (1997, 1989). These provide distinct and related channels through which people make/practice leisure. Each of these may have a particular consequence for space in which practice takes place. Ritual enactments and artistic production in leisure may entail very distinctive spatial practices. They may also be considered as embodied practices.

The idea of social engagement as informing the meaning, perhaps especially through the sharing of practice and meanings, is significant but largely undeveloped. The possibility of embodiment in leisure practice emerged in terms of carnival (Jackson 1988). Using Bhaktin's insight into liminality, carnival offers a very significant example of the semi-detachment of the subject from conventions. Through the body's disengagement the subject feels able to express the self. It may be that in other, perhaps more commodity-controlled leisure spaces with regulation in the form of social conventions and semiotics such liminality is less easy. Conversely, the constitution of, for example, imaginary, fairytale, sexually disarming or age-suspending spaces may provide opportunities for such liminality, as the interpretation of Brighton's dirty weekends suggests (Shields 1991). However, Brighton was semiotically constituted through a complexity of social, bodily and other material, and the crucial metaphors seem often to be triggered in the spaces of practice between rather than so much in carefully crafted representations. Spaces for liminality can be constituted through practice as the examples of gambling, wandering in unfamiliar landscapes, and drug-taking have been used to demonstrate (Cohen and Taylor 1993).

With reference to an especially artistic practice, that also is very deliberately about both the body and the expressive use of the body, Radley and then Thrift have developed a discourse on embodied expressivity through a consideration of dancing (Radley 1995, 1996; Thrift 1997). It is likely that the limits of embodied expressivity are not exhausted in dancing, nor the reaches of artistic enactment in what are properly known as the arts, and such an approach deserves much greater attention in considering all kinds of leisure practice.

Embodiment, sociality, ritual enactment and artistic production can all make use of the materiality of space in their spatial practice. The importance of materiality in leisure spatial practice is a theme explored in depth by Mike Crang in the form of using photographs in social practice rather than seeing photographs as only visual representations (1997, this volume). Moorhouse points to the importance of materiality in leisure as the minutiae of social engagement, for sharing enthusiasm, making friendship and even community (1992). He refutes the dominance of leisure metaphors (the phallic prowess of motor-racing) and argues for a greater emphasis on what sense people are making in their exchange of feeling and lay knowledge in the practice. Radley argues the significance of material artefacts in the constitution of memory (1990). Of course, materiality and metaphor intercut (Keith and Pile 1993). What is very little explored is the way in which this intercutting works, whether commodified, consumption-oriented or not, and the most pertinent material so far is in terms of shopping (Miller *et al.* 1998).

In this chapter I consider empirical evidence of the content of two leisure practices to try to enlarge these various themes in terms of spatial practice. My emphasis is on the embodied subject. Immediately it is necessary to clarify that the subject is understood to be profoundly socialised (as argued cogently by Wilson (1990) and by Bourdieu (1990), on gender and on class), and that contexts within which she practises space are highly encultured and socialised. Young modifies her stronger claims to this in a later introduction (1990: 11). These components are not ignored in this account. However, this account seeks to focus these issues through contemporary lay geographies, reflecting Crang's work on the spatial practices of work and his ideas on the possible field of lay geographies (1994, 1996). Throughout this work rages debate concerning the encounter of practice and context, the socialised subject and the practising subject, socialised contexts.

Using two examples of leisure practices this chapter makes an effort to understand how people make spatial practices within this collision of influences, as socialised subjects, using culturally and socially contextualised spaces. The two examples this chapter considers are rally caravanning and allotment holding (plotting). Like any other, these are distinctive leisure practices happening in very distinctive spaces. Plotting takes place predominantly in the 'city', the other in the 'countryside', typically on the edge of a village or away from settlement in farming or open country. Urban and rural, these spaces are heavily contextualised metaphorically and materially. City and country cultural geographical metaphors that remain undiminished and are inflected with ideas of nature. These keywords work in relation to particular commodified representations and contexts and popular metaphors, and each practice and its peculiar space is surrounded by its own metaphors in popular culture.

It is because both of these examples unsettle familiar and perhaps dominant versions of leisure, popular geography, city, country, that these provide us with an interesting means of exploring those prevailing metaphors and meanings, and how they work and may be reworked in lay practice and knowledge (Creswell 1996). One example, plotting, is not commodified and people rent their plot of ground from local councils or private landowners, although tools and other equipment can be expensive. A caravan usually costs about £18,000 but there is a strong second-hand market. Sites used for weekend rallies are not expensive although many rally caravanners also use highly commercialised sites that can accommodate hundreds or thousands of vans at different market levels. Allotment holding has a strong tradition of working-class self-help, and rally caravanning has a heritage from camping traditions and enjoying the outdoors, in others' company and has included different class groups over many decades (Crouch 1997).

Popular metaphors of allotments are constituted of particular kinds of people, retired, working class, men, a particular cultural capital (Crouch and Ward 1997). Caravanning's popular culture was epitomised in the comedy film 'Carry on Behind' (1972), which depicted caravanners as working-class people insufficiently coping with somewhat increased wealth and not coping with their relationships. The spaces with which these practices are associated have a similar cultural constitution, figured as run down landscapes, and bulky vans with people behaving too loudly for traditional 'country leisure' (Crouch and Ravenscroft 1995, Crouch 1997).

Fields of practice

I have been fascinated by what these two practices may mean to people who do them, because they would seem to fit uneasily with increasingly lifestylised leisure, but paradoxically exhibit significant aspects of lifestylisation. In order to try and make better sense of these practices, and to try and inform this paradox qualitative empirical case studies were made of each. Their ethnographic investigation included participant observation and long semi-structured interviews, a series of informal discussion groups each over the duration of a weekend, supported by archival reference.[1] In this chapter empirical material is quoted in two ways: as interviewee quotes and as field diary notes. The two case studies are discussed together, but not in a closely comparative way. Both cases are used to explore aspects of spatial practice. However, distinctive facets are drawn out and considered in order to enrich insight into leisure's lay knowledge.

People rent allotments to do what other people may do in their garden, cultivate ground, grow flowers and food as exercise, as aesthetics, as an exercise in love and care (Crouch and Ward 1997). An allotment is separated by up to a mile from home, and one 'plot' is situated amongst dozens, even hundreds of others. This confers particular characteristics on the significance of the space and the practice that happens in it, for example, that people share a communal space, whether or not that may be welcome. Sites are fringed by houses, other buildings, parks or other 'open' spaces. The plots are usually arranged serially in parallel rows of roughly fourteen per acre.

Recreational caravanning in the UK takes numerous forms. Two thirds of a million households are members of the two main clubs in the UK (Camping and Caravanning Club, Caravan Club) and four million holidays are spent caravan-ning each year (House of Commons 1995). About twenty per cent of the membership of the main clubs takes part in weekend club rallies. Between ten and fifty households attend a caravan rally, usually in a field on a farm, in a space behind a pub or a garden centre, anything from one to one hundred miles from members' homes. A van is taken to a site where activities are arranged amongst members and much of the time is spent informally alone or in company. Vans are likely to be arranged in rows or groups, limited by safety regulations.

Plotting entails prolonged investment of human energy/care, whereas rally caravanning usually involves a different site each weekend, and many people attend a dozen or more weekends a year. Different ways of relating to space may be involved in these different time practices.

What people do varies considerably. Interviewed caravanners listed walking and strolling (28 per cent), one-fifth sew, fish, play bingo or do another hobby whilst on rallies. Somewhat fewer visit local sites, towns and events, and many included the journey to the site as an important part of the event. They listed fifty differ-ent activities, from cycling to shopping, playing games, sitting in chairs, blackberry picking, folk music, car boot sales, four-wheel drive watching. Nine per cent noted 'enjoying the countryside' which for many merges with walking, 20 per cent liked 'doing nothing'. Half those interviewed noted 'sharing time' and 'doing things together'; 20 per cent valued 'socialising' and social drinking. Exactly what these diverse activities mean is explored in the main part of this chapter.

A quantitative survey of allotment holders identified their main activities as those connected directly with working the ground, looking after plants and general cultivation, repairing built structures, keeping the site tidy, sometimes extended to include getting exercise (Saunders 1993). Qualitative research gives voice also to the importance of social activities and aesthetic productivity (Crouch 1992, Crouch and Ward 1997).

Embodied space, embodied practice

Participating in a caravanning weekend or in allotment holding is to participate in complex activities. A first observation is that people move around, space seems to be encountered in numerous, subtle ways. In caravanning, walking across the field I am surrounded by activity, children playing to the left down the slope, adults chatting along the other side. People pass me jogging, toddlers try and ride bikes ahead, in the distance people walk their dogs and return to the site from somewhere else. Many people are lying down or sitting by their vans as I walk the greens between groups of vans. I walk over to join the group I am with, aware of the activity behind me.

(DC fieldnotes Suffolk 1997)

People get together in barbecues in spaces that can symbolise group identity; at special 'flag ceremonies' club announcements are made and prizes given; people leave the site and return, play games, have coffee together in multiple house-hold groups.

Plotholding involves a range of activities from cultivation (itself a very complex practice) to moving bulk materials, repairing sheds and plant shelters, maintaining shared spaces such as lanes between plots, talking with other people, collecting up materials and sometimes giving them away (Crouch and Ward 1997). Sometimes plotters express vividly an explicitly embodied awareness of their practice:

Working outdoors feels much better for your body somehow . . . more vigorous than day to day housework, and more variety and stimulus. The air is always different and alerts the skin . . . unexpected scents brought by breezes. Only when on your hands and knees do you notice insects and other small wonders . . . my allot-ment is of central importance to my life, I feel strongly that everyone should have access to land, to establish a close relationship to the earth, . . . essential as our surroundings become more artificial.

(Karen, Durham)

This observation connects with the tradition of land claims and self help amongst the working class that precipitated allotments in the UK (Crouch and Ward 1997).

In each case the subject is engaging space in an embodied way, reaching and knowing, 'making sense of' the surrounding space through tactile and other-sensual experience. These spaces, sites and plots, are embodied through the practices, doing things. The space is a surrounding volume, it is a multidimensional practice where the subject is alert in different degrees and ways to an encounter made up of numerous fragments as a patina of multi-sensual body-space explored at length by Merleau-Ponty (1962). This provides the subject a very different access to making sense of the world than merely a flat-view gaze (Urry 1990; Cloke and Perkins 1998).

In this there is an interesting connection with nature that appears also to be embodied. Through being close to the ground the plotter confronts new

263

dimensions of space and feels this gives access to a different kind of encounter than just looking. Through sitting, playing games, making barbecues, people caravanning can discover 'nature' or 'landscape' as a series of different surroundings and broken volumes encountered through doing things, meeting people, preparing a barbecue, cultivating. This may be despite their not prioritising 'nature' as part of what they do, frequently felt only as a sensual backdrop for other events. However, it matters in terms of what they feel they are doing:

> No, it's not the countryside that matters so much, but if you went there just for the day you wouldn't notice so much . . . there was a place up the road, just a horrible little field. Last week was totally different, a pretty village, deer walking round, forest all round us. We took bikes and went for a cycle. I wouldn't have known the place existed only a few miles from home. You can enjoy yourself. I don't think you have to travel a great distance from home to get a different scenery.
>
> (Julie, Essex)

There is an implicit paradox here that points to the informing significance of space realised through these other activities.

For plotholders, there is a metaphorical inversion exemplified in the example of a young couple who live in the city and visit their plot on the edge of a large site that is fringed by a busy road and terraced housing:

> We come here for the peace and quiet. On a nice summer's day you feel like you're in the country, it's like being in the countryside, in the city.
>
> (Linda, Birmingham)

Whilst it is evident that plotters work with abstract ideas of what country, city and nature may be like those ideas are encountered by working with (or against) nature in cultivation. In Karen's case this was combined with a strong ideology of 'land' that she felt she sustained in practising space itself, but was also surprised by the close encounters of practice itself that she drew together with a more intimate knowledge.

Simple tasks in caravanning are remarked in relation to material features of the space, in this case in terms of a regular walk from the caravan site across a river bridge. John, 70-something, talks with his wife, Betty:

> for me it's the wide open spaces that attract me up here. You get out for a walk now and again – do you remember that little river in spate during that storm when you went across to bingo? It was a terrible day, it was flooded.
>
> (John, Co. Durham 1997)

'Nature' is related through experience, practice: 'You're just closer to things in a more natural way.' In the case of plotting there is a clear objective in terms

Plate 8 SLUG: San Francisco League of Urban Gardeners

of nature that can be unsettled or sustained in new ways in practice, but in caravanning this is more accidental. The body and sensual feeling enter language in unexpected ways: 'You feel you are breathing air into your living'; 'When you are caravanning you can let your hair down'. This is an expression used by plotters too. This suggests an awareness of bodily participation and an embodied response to this particular space *and* practice, the body not alert according to particular regimes of regulation – although both have regimes of clubs and landowners (Crouch and Ward 1997) – but felt to be freely participated. Especially in caravanning people talk of 'slumping down in a chair', the physical enjoyment of walking and a different physical and sensual pleasure of 'mooching around' (moving around getting a sense of the space, its artefacts and what other people are doing).

You sit back, that's the reason I go caravanning, but I also enjoy long walks.

(Jon, Essex 1997)

Space is used to transform the way of making sense of being somewhere and doing something chosen on one's own terms.

This self-determination appears in another way. Caravanners frequently remark on the attraction of being able to 'go where you like when you like'. This may be over-imaginative and is informed by a tradition, an ideal of camping and caravanning, sustained in contemporary club magazines, of reaching the 'outback', of countryside as escape. Yet practical limitations on movement and landowners' restrictions on parking 'anywhere' are significant. However, this desire is practised through choosing different sites to visit on different weekends, going with people one chooses. This produces a partly abstracted knowledge of freedom, where different sites merge in the idea of practice, but the abstraction is itself practised through repeated rallies where embodied knowledge is shared and added and refigures the abstract idea. Plotholders on the other hand invest themselves in particular spaces embodied with productive and imaginative practice 'creation, nature, my allotment means everything to me' (Delia, Northampton, 1994, in Crouch and Ward 1997).

The space as surrounding the body is very important to caravanners: 'in the middle of a field, in the middle of nowhere',

here you sit out in weather you wouldn't dream of doing at home, wear fewer clothes, people behave differently.

(Doris, retired)

The physical encounter with the materiality of the site is felt to be metaphorical of the friendship caravanners value amongst the group, the relaxed body that is also rediscovered through practice. Embodiment and body-space is also social:

Here you're not on your own but surrounded by other people. We have always tried to do our own thing during the day, like walking, and meet up at night-time. It's like coming back to a temporary village.

(Denis, Yorkshire, 40s)

Space is something that people feel around them, not as demarcated by club regulation plots and required distances between vans but by the materiality of the site, trees, other people's paraphernalia, van decorations, flags collected on previous rallies that show where you've been, resembling Moorhouse's remarks on the material minutiae of social enthusiasms (1992). They move around the barbecue and relax, and wander over to the flag ceremony. The space in which people gather for a caravan rally is marked by a group, or rows of vans and serves as a metaphor of friendship, enacted through a number of activities in which people share. Barbecues are perhaps the most significant spontaneous event, although many clubs regularly arrange quizzes or other games to bring people together.

People get to know a caravan site through movements and also through restive reflection, 'doing nothing', 'just lazing in a deckchair with a pint of beer'. Whilst people stop and stare, or gaze, they also move around, talk to each other, play with their children. Those who spend time *plotting* not only invest their interest in cultivating ground and plants, digging and turning, pulling and bending, but walk across the site, meet other people for a chat, collect water, and survey the scene. The plot and the site are learnt and become habitually familiar through what the body does and through senses as well as social engagement. Not only does this seem to represent an active multi-sensuality in the encounter with space, but in each case people are expressive.

Ironically, this is perhaps more obvious in terms of cultivating land, which may seem too much hard work. However, it expresses learnt skill, love and care, sensitivity, and an everyday aesthetics. Cultivating ground becomes important in self-identity and the result, whether that is harvested crops or well-tilled land, becomes expressive of that identity. The artefacts of cultivation become artefacts of expression that is also intersubjective, as friendship, a gift relationship,

If you give somebody anything, they say – where did you get it from – I say I grew it myself. You feel proud in yourself that you grows it.

(Sylvester, Birmingham 1994)

The body tests its physical boundaries and fulfils its own desires in movement, orders and unsettles, relates and punctuates its space through motions, the way the body makes out space, places itself, incorporates its sensuality. Walking across these spaces, like the woman on her way to bingo, resting in a particular perhaps favoured, position, encounters knowledge of people and events and expresses the encounter with them. Practising space like this contributes to the feeling that makes the leisure important and enhances the pleasurable feeling of doing it. Radley's (1995, 1996) work on expressive practice and Finnegan's (1997) on artistic production have been noted already, in relation to dance and to music. It would seem likely that there are less obviously expressive leisure practices that share some of these constituents, work alongside rules in the use of a plot in ways of cultivation, in recycling, in positioning a van and in arranging games and flags.

Many people who have an allotment do so in order to be solitary, reflective, meditative. Some caravanners also seek escape from human contact. However,

267

Plate 9 Peaceable Kingdoms

in each case we encountered numerous households who may have been seeking to escape particular social contexts of work and locality, but who emphasised the importance of friendship in this leisure. Each practice includes rituals and through these and more informal impromptu encounters people explore social relationships. This social activity presents another way in which people under-stand and come to value the peculiar surroundings of allotment and caravan sites. Body space is shared as people work their ground and as caravanners look after their van, get together for tea or become aware of other people doing the same, often through seemingly trivial acknowledgement of each other.

> *Here you've got something that's almost a temporary village — or is that too romantic? It is a thing to escape to, really. So for a lot of people it's not the same as at home, it's a bit of a sanctuary really. I don't think people started camping or caravanning for that sanctuary, it probably comes out of what you're doing because of what you're doing and of values you treasure and how you respect other people, their space.*
>
> (Denis, 40s)

Plotters talking suggest a similar alertness to body space and to practice:

JOHN: I started at sixty-five and I'm seventy-five May gone. I came and found her here. We learn things from each other. You are very social and you are very kind. You make me feel good, you don't come and call at me. Things from your garden you give me, little fruits, which I have enjoyed ever so much.

ALLEN: And I've learnt from him, learnt real skills about planting, Jamaican ways of growing and cooking; learnt about patience and goodness and religion too, it all links in.

She is retired and around sixty, he is a decade older; each expresses a friend-ship, largely experienced in proximity of what they are doing as well as in verbal exchange and social activities, but 'simply digging' and cultivation emerges through this example as a social/sociable activity. Friendship can be a major event in caravanning.

> *I come because Sarah's coming, so that we have a social evening away from home. You can sit in the corner and say — sod the social life — we always go to the social evenings, a give-and-take part of it. . . . It's socialising isn't it, that's what I enjoy, the fact that you can relax and unwind without the hum-drumness of being at home, without the ties.*
>
> (Julie, Essex 1996)

These artefacts, the places where the barbecues are made and the part of the allotment site where people gather to talk, are embedded in human friendship. There is an underlying distinction of gender in the way that two of the cara-vanners express practice. The male respondent identifies the almost ideological importance of the imagined community (temporary village, sanctuary), roman-ticises, and the woman prioritises the chance of being with a close friend from home at a more intimate level than usually found possible in everyday pressures and represented in simple, important artefacts.

> *I started putting ornaments in because of Sarah. She kept saying fancy coming here with a caravan (they live only three miles from the site). She dresses hers up with ornaments, it became a bit of a joke, rivalry. Bought a lamp for £3. It was a joke, a bit of a mickey-take. You dress it up like at home, you take the dressings with you.*
>
> (Julie, Essex 1996)

In each of these examples sociality seems to be more than a function of gender and class composition. Identity is constituted of club membership or renting a plot respectively, that carries social and cultural identity. It is also constituted of shared activities and shared space that comprises particular, familiar artefacts, friendships. Sociality is more than convenience and engages these two dimensions of practice and 'membership', that confers an infective, shared enthusiasm (Moorhouse 1992).

Refiguring processes

Through practices, some of which may be formalised in the word ritual, people make sense of the self and go beyond the self in engaging emotion and social friendship to 'the creation of non-commodity values', Gorz's rare expression of the human subject (1982: 81). These subjective qualities are very important in leisure practices, and constitute significantly what leisure practices are, as discussed below. Urry (1995) and Finnegan (1997) as well as Bishop and Hoggett (1986) have all referred to Gorz.

These kinds of practice may not always be easily contained in a notion of 'artistic creation' (Finnegan 1997) but these subjects are participating in making friendships, in making activities, if through very humble, seemingly trivial events and materials. In each of these cases friendship, love and care are practised in numerous activities. This happens in making barbecues, doing games, helping each other out, for example, when there is a problem with the caravan stuck in the mud; sharing seeds and swapping hints on cultivation, helping clear each other's plot and tackling resistant weeds (Crouch and Ward 1997). Friendship like this can be both embodied and spatial. It is embodied because it is developed and sustained through what the body does rather than in an abstraction, and makes particular use of space. People being physically together, sharing activities, the body becomes aware of a shared body space that is also a social space. Sociality is also located in particular spaces demarcated, practised in the course of expressing care, sharing enthusiasm.

In these spaces people can feel safe to explore imagination, to experiment, to be reflexive (for some 'at home', but also away from 'hum-drumness') and to be creative, and to simply feel able to 'do nothing', which probably means to discover the self. This includes to be 'somewhere else',

> in your little island of the allotment, with this shed place where you pretended you lived.
>
> (Tom, Birmingham 1994)

often implying to be *someone* else. To be somewhere else is to engage the imagination, as is the idea from caravanning of being 'where you like when you like'. Obviously the reality of where/when is limited, but through this come other possibilities of getting in touch with the self, not of total freedom but a liminality of experimentation through self and space, a reflexive imagination, refiguring space and signification (Lash and Urry 1994: 50). This echoes Kristeva (1996). Reflexivity can give collective solidarity. In both our examples this imagination is embodied.

> Caravanning, it makes me smile inside. I mean, everyone comes down to the ford and just stands there and watches it, watches life go by. It's amazing how you can have pleasure from something like that.

(Jim, Weardale 1997)

The activity and its space are enlarged in the imagination.

There are elements of the poetic in these practices, of transformations of the self that utilise embodied space through imaginative practices. In the varied ways we have observed through these cases the subjects are constantly refiguring the spaces they use (Wearing and Wearing 1995). This is different from the outer limits of de Certeau's guerrilla tactics and 'lucky hits', but transforms what a space means nonetheless. The space no longer conforms, for those caravanning or allotment holding, to regulated space, and even if their practice *is* regulated in several ways, they indulge in other meanings. Whilst their practice may be informed by socialisation their practice goes beyond the dimensions of these limits.

The allotment becomes a site of close social friendship and where structures can be produced that go further than the needs of functionality and border on the romantic/expressive. The caravan site can be envisaged as a mobile village; the field becomes a metaphor of practice:

> in the middle of the field there is a group of vans, they're all my mates, we have a barbecue and good booze.

(Jon, Essex 1997)

The spaces, the fields and plots, are both imagined and felt in doing, with 'imaginary powers of feeling' (Thrift 1997), around the barbecue and around the corral of 'vans'; in allotments between the individual plot and the whole array of plots across a site, in the creative production of self-built sheds and other contraptions. Enclosure becomes a metaphor of friendship, although as others expressed to us this can be claustrophobia. That is both spatial and social, strongly embodied.

Ideas of the self and of social relations are figured and refigured through practices. They are practised, as a process of negotiations and adjustment. These practices unsettle ideas of the self, of time and space. They are also performed according to particular situations of time and space and the meanings such configurations of time and space make possible. Chaotic and complex identities can be managed/enjoyed perhaps by being 'held' in lifestyles that make sense, if sometimes uneasily.

Through many flashes of imaginative practice these spaces and practices keep alive possibilities of the self, embodied and social. These practices of imagination can relate to other arenas of life practices or make them more tolerable. Imagined communities are felt to be 'located' socially and spatially through practice. Leisure practice can unsettle and be used to intersect imagination, metaphor, sociality, and embodiment. The space becomes important in the imaginative realisation of the self and friendship. There is constant adjustment and negotiation, grasping space. The body is managed and there is social management, but both held and played with reflexively. The practices can happen liminally. Knowledge through bodily adjustment close to the ground, in particular body-related spaces and in doing things, transforms regulated practices and spaces. Plot-shapes and rows of vans are no longer a series of regulated rows and become something else in imaginative practised embodied knowledge, constituting a knowledge of doing.

Conclusions

This chapter has sought to articulate leisure practice and knowledge through the subject. This has tended to foreground the subject and therefore provides an imperfectly balanced argument. However, the lack of close investigation of practice has meant that the burden of much debate has been the other way. It is important to develop insight into more practices and their closer relationship with context and socialisation.

Recent popular map-making provides a means of understanding the imaginative as well as practical reading and use of space (Crouch and Matless 1996), refiguring metaphors through material contents and practices. A similar refiguring of knowledge is happening in these leisure practices, encountering space and practice itself. Practice refigures metaphors of spaces and practices. Both de Certeau (1984) and Radley (1995) in different ways acknowledge the positioning through practice of artefacts. In these cases, artefacts include the spaces themselves, their arrangement, content, signified in practice. Reflecting on these two case studies it is possible to disentangle some of the complex and inevitably frayed threads of meaning, practice, ontology, lay knowledge, that shift across practices like these. This includes the geographical-cultural complexity of how this works as one engages the active, reflexive subject: doing things, thinking things, influenced by things.

What seems to emerge from this is a complexity of constituents and influences. Giddens argues that contemporary human practice frequently seeks to grasp security from all of this (1992). Leisure/tourism is sometimes felt to be a useful arena or even tool through which security is possible (Lash and Urry 1994). Whilst the constituents of practice discussed here may be considered in terms of self-generated leisure practice (Crouch and Tomlinson 1994), focusing

Plate 10 Ontology/practice

Photo: Richard Grassick

the subject suggests that commodification may not provide the break often considered in terms of leisure practices. In both of these cases commodification is ever present, if by default. Many contemporary plotters who are not 'working class' define their attraction in terms of an ideology or a set of ideas. These grasp issues of land, cultivation, qualities of human relatedness they feel are also shared with earlier generations of plotters: 'I come down here to get away from the commercial world' (James, Bingley 1992). Numerous rally caravanners typically said 'I always avoid commercial sites' (Bernard, Essex 1994). Many of them also found commercial sites a very different kind of leisure practice. How far, however, commercial sites radically change the complexity of embodiment, sociality and even imaginative practice remains unclear, and it may be that people simply prefer more facilities sometimes, being with different people. Instead the embodied subjectivity that emerges in unlikely leisure practices suggests that the practice of theme parks, historical enactments and being a spectator at golf may resemble those observed in these cases more than may be imagined. These may be in very different combinations, socialised forms and so on, but more complex than even the most recent insights into consumption suggest.

Our two cases exhibit an aesthetic productivity that stretches beyond likely aesthetic endeavours. They are also expressive. Allotment holding produces both metaphorical and material aesthetic production including an intricate landscape (Crouch 1992a). Caravanning is similar, although its material aesthetics is of a brief experienced time trajectory (the paraphernalia of vans, chairs, flags, numerous bodies and so on). Both examples figure and refigure metaphors of city, country, landscape and nature. Each utilise metaphors like these in figuring the practice, but through that practice these metaphors are refigured. The opportunity for self expression, self exploration, creativity, alternative socialities and becoming aware of the world through a liberated sense of embodiment refigure each of these. Each involves utopian ideas connected with the rural and nature, if muted. We make play with these diverse components.

Lay knowledge is figured through these practices. At particular times these practices become important in adjusting, coping, negotiating other components of life practices. At particular times these practices are important because they offer a rediscovery or remaking of identity, expression of love and care, liberate regulated bodies and exercise the imagination.

When we participate in leisure practice we refigure places and practices into new knowledge, however brief. We have a chance to refigure our knowledge of ourselves. Lay geographical knowledge is a kaleidoscope, a fractured, incomplete, uneven compilation yet one that people seek to resolve or unsettle. Doing leisure can unsettle other constellations of knowledge, and may be a reason for doing it. It can adjust the tension between socialised contexts and the embodied practice. People can enlarge their knowledge mutually of particular places, of themselves and others through practice.

Note

1 Forty rally caravanning households were interviewed between 1996 and 1997, with additional material from 1994 and 1995. Over sixty allotment households have now been interviewed since 1986. Throughout this time, in both studies respectively, participant observation has been undertaken, as well as extensive visual ethnography. The author acknowledges that he is also now an allotments advocate, including both providing expert evidence to the House of Commons (1998) and working on several TV films 1992–1996.

Acknowledgement

A large part of the interviews and fieldwork in this chapter (caravanning) was undertaken by Adam Colin. Anglia University supported this research project and The Leverhulme Trust supported the Allotment work, which I did with the friendly insight of anarchist social historian Colin Ward.

References

Bishop J. and Hoggett P. (1986) *Organising around Enthusiasms: patterns of mutual aid in leisure*, Comedia Publishing Group: London.

Bourdieu P. (1984) *Distinction: a social critique of the judgement of taste*, trans. R. Nice, Routledge: London.

Bourdieu P. (1990) From Rules to Strategies: an interview with Pierre Bourdieu, *Cultural Anthropology* 1(1): 110–20.

Chaney D. (1996) *Lifestyles*, Routledge: London.

Cloke P. and Perkins H. (1998) Cracking the Canyon with the Awesome Foursome: representations of adventure in New Zealand, *Environment and Panning D: Society and Space* 16.

Cohen P. and Taylor L. (1993) *Escape Attempts: the theory and practice of resistance in everyday life*, Routledge: London.

Crang M. (1997) Picturing Practices: research through the tourist gaze, *Progress in Human Geography* 21(3): 359–73.

Crang P. (1994) It's Showtime: on the workplace geographies of display in a restaurant in southeast England, *Environment and Planning D: Society and Space* 12: 675–704.

Crang P. (1996) Popular Geographies: guest editorial, *Environment and Planning D: Society and Space* 14: 1.

Creswell T. (1996) *In Place/Out of Place*, Minnesota University Press: Minnesota.

Crouch D. (1992) Popular Culture and What We Make of the Rural, *Journal of Rural Studies*, 8(3): 229–40.

Crouch D. (1992a): Landscapes of Ordinary People, *Landscape*, Berkeley, 31 March: 3–8.

Crouch D. (1997) 'Others' in the Rural: leisure practices and geographical knowledge, in P. Milbourne (ed.), *Revealing Rural Others*, Cassell: London.

Crouch D. (1998) The Street in the Making of Popular Geographical Knowledge, in N. Fyfe (ed.), *Images of the Street,* Routledge: London.

Crouch D. and Matless D. (1996) Refiguring Geography: the parish maps of common ground, *Transactions of the Institute of British Geographers.*.

Crouch D. and Ravenscroft N. (1995) Culture, Social Difference and the Leisure Experience: the example of consuming countryside, in G. McFee *et al.* (eds), *Leisure Cultures: values, genders, lifestyles*, Leisure Studies Association: Brighton..

Crouch D. and Ward C. (1997) *The Allotment: its landscape and culture*, Faber and Faber/Five Leaves Press: London/Nottingham.

de Certeau M. (1984) *The Practice of Everyday Life*, University of California Press: Berkeley.

Dickens P. (1996) *Reconstructing Nature: alienation, emancipation and the division of labour,* Routledge: London.

Finnegan R. (1989) *The Hidden Musicians: music making in an English Town*, Open University Press: Buckingham.

Finnegan R. (1997) Music, Performance and Enactment, in G. Mackay (ed.) *Consumption and Everyday Life*, Sage: London.

Game A. (1991) *Undoing Sociology: towards a deconstructive sociology*, Open University Press: Buckingham.

Giddens A. (1992) *Modernity and Self-Identity*, Polity: Cambridge.

Gorz A. (1982) *Farewell to the Working Class*, Pluto: London.

Gorz A. (1984) *Paths to Paradise*, Pluto Press: London.

Gregson N. and Crewe L. (1997) The Bargain, the Knowledge and the Spectacle: making sense of consumption in the space of the car-boot fair, *Environment and Planning D: Society and Space* 15: 87–112.

House of Commons Environment Committee (1995) *Fourth Report*, vol. 1, HMSO: London.

Jackson P. (1988) Street Life: the politics of carnival, *Environment and Planning D: Society and Space* 6: 213–27.

Kayser Nielsen N. (1995) The Stadium in the City: a modern story, in J. Bale (ed.), *The Stadium and the City*, Keele University Press: Keele.

Keith M. and Pile S. (eds) (1993) *Spaces of Identity*, Routledge: London.

Kristeva J. (1996) *The Portable Kristeva*, Columbia University Press: New York.

Lash and Urry (1994) *Economies of Signs and Space*, Sage: London.

Ley D. and Olds K. (1988) Landscape as Spectacle: world's fairs and the culture of heroic consumption, *Environment and Planning D: Society and Space* 6: 191–212.

MacNaghten P. and Urry J. (1998) Constructing the Countryside and the Passive Body, in C. Brackenridge (ed.) *Body Matters: leisure images and lifetyles*, Leisure Studies Association: Brighton.

Merleau-Ponty M. (1962) *The Phenomenology of Perception*, trans. R. Nice, Routledge & Kegan Paul: London.

Moorhouse R. (1992) *Driving Ambitions*, Manchester University Press: Manchester.

Miller D. (1987) *Material Culture and Mass Consumption*, Blackwell: Oxford.

Miller D. *et al.* (1998) *Shopping, Place and Identity*, Routledge: London.

Radley A. (1990) Artefacts, Memory and a Sense of the Past, in D. Middleton and D. Edwards (eds), *Collective Remembering*, Sage: London.

Radley A. (1995) The Elusory Body and Social Constructionist Theory, *Body and Society* 1(2): 3–24.

Radley A. (1996) The Triumph of Narrative? A reply to Arthur Frank, *Body and Society* 3(3): 93–101.

Rose G. (1993) Geography as a Science of Observation: the landscape, the gaze and masculinity, in Felix Driver and Gillian Rose (eds), *Nature and Science: Essays in the History of Geographical Knowledge, Historical Geography Research Series*, vol. 28, Feb.

Saunders P. (1993) *Allotments 2000*, National Society of Allotments and Leisure Gardeners: Corby.

Shields R. (1991) *Places on the Margins*, Routledge: London.

Thrift N. (1997) The Still Point: resistance, expressive embodiment and dance, in S. Pile and M. Keith (eds) *Geographies of Resistance*, Routledge: London.

Urry J. (1990) *The Tourist Gaze*, Sage: London.

Urry J. (1995) *Consuming Places*, Routledge: London.

Warren S. (1993) This Heaven Gives Me Migraines: the problems and promise of land-scapes of leisure, in J. Duncan and D. Ley (eds), *Place / Culture / Representation*, Routledge: London.

Wearing B. and and Wearing S. (1996) Refocusing the Tourist Experience: the flaneur and the choraster, *Leisure Studies* 15: 229–43.

Young I. M. (1990) *Throwing Like a Girl and Other Essays in Feminist Philosophy and Social Theory*, Indiana University Press: Indiana.

19

KNOWLEDGE BY DOING

Home and identity in a bodily perspective

*Niels Kayser Nielsen**

Introduction

Today there is general agreement that no matter whether one distinguishes between group or individual identity, identity is more attached to the dimension of space than to the dimension of time. This is also what our everyday language tells us: we are dealing with home place but not with home time (Morley and Robbins 1993).

There also seems to be a growing understanding that the question of identity and space is related to the distinct feature of space, in contrast to time, that you, deliberately or not, move in space, in the phenomenological meaning of the word, whereas this is impossible in time. Time is a much more absolute condition than space. Time is irreversible; space is not. Space is a 'working partner' with whom you interact in order to create new aspects of life. Space stands in contrast to time, where the only possibility will be to make the best out of it. Time cannot be improved; space certainly can. So the first intention in this chapter is to explore the influence of movement on making an identity.

On the other hand, it also seems that this freedom to 'improve' space is only possible when outside the 'realm of necessity'; i.e. when you are not working, resigning yourself to a certain task and a certain aimed rationality. Consequently, the freedom of improvement of space is closely connected with leisure; as one is released from what Max Weber calls aimed rationality, the possibility of gaining knowledge of yourself is provided in leisure time. Leisure, as a contrast to work's realm of necessity, is the realm of contingency, i.e. a sphere of unpredictability and openmindedness. Therefore, the second intention of this chapter will be to point at the potentials of leisure for making knowledge of yourself possible

---

* This chapter has been translated from Danish by Ann Marlene Lau.

through knowledge by doing, the result of which is what has, a bit vehemently been called 'visceral' knowledge (Berman 1990: 134). Nevertheless, this bold statement refers to the well known, although often hidden, circumstance, that human knowledge is made up of the body's inner perceptions of boredom, fun, sexuality, humour, offences, crying, sneezing, anxiety, addiction, temptation, running, falling, standing (Berman 1990: 116).

However, it is important to underline the fact that this new knowledge of oneself is not 'delivered'; nor does it come out of nowhere. It is a question of gaining it and the only possibility of gaining knowledge in relation to space is through moving and/or sensory perceptions. So, gaining knowledge through space is indispensably connected with your body and its movements and perceptions. But it is just as important to underline the fact that if this knowledge gained through moving and perceiving is to result in a sense of identity, the movements cannot take place in an abstract space. The movements must take place in a space that it is possible to appropriate as a kind of balancing act between one's movements and certain fixed points and routes in space (Bausinger 1980: 27). So, today when identity is no longer given through work but tends to be constructed, and constructed more through leisure than work, it still seems to be a condition that identity is related to doing; i.e. by moving and living in a space that you appropriate by your body and not by your thinking. Identity is intimately connected with a space that you make your own by moving in it. Identity is something gained in the spatial dimension, enacted in leisure time as a result of bodily experiences of both being surrendered to space and being able to form and create space by your movements in it. In short, bodily activities in space are not a practice 'standing for your identity'. On the contrary, your identity is produced by the experience of certain distinct and idiosyncratic bodily actions and, today, the use of the body is more related to leisure than to work. In contemporary society, most people do not use the body at work, as much as in leisure time. This is important since people do not act as something; they become something by acting (Kayser Nielsen 1997b). Here it is appropriate, however, to define what is understood by space.

The concept of space is, so far, as abstract in itself as the concept of time. Therefore it seems too dim and not specific enough to constitute an anchoring of identity. So we must include the category of place which is concrete and specific. Place may smell and sound, space never does. Compared to the meaning of the word place, the word space often denotes the meaning of empty space, particularly if, in connection with the concept of place, we also think of the way we relate to a place as being the identity maker – and marker. Cultural creation is important in this process, a projecting as well as a receiving relation to a specific place. In other words, we must, at the least, distinguish between space and place.

Space and place

One step further makes another distinction, between three kinds of attachment to a place: as birthplace, as place of experience and place as history. Of these three definitions of place, place as birthplace is the most abstract and irreversible aspect. Being born infers being born at a specific place which states a condition and a certain affiliation, but apart from that it does not seem very identity forming. The fact that a person has been born is not indicative of what that person has been doing. Therefore there is a need to practise place. Leisure provides the possibility of changing space into place through a multiple sensory range of experiences and through seeking, deliberately or not, certain landscapes and places, creating and appropriating them by moving and perceiving in them. This means that place is seen from a relational point of view. In connection with the concept of birthplace only a mark of condition is stated. It merely shows that by virtue of our birth we occupy a certain place in the world.

On the other hand, in stressing place as a place of relations we mark the activities and social relations, particularly through non-constrained leisure activities that we have made and still take part in. The body as a condition is not in focus here, but it is rather the relations of the body, i.e. its obligingness and its ability to act: run, jump, make love and fight which is the essence of practice. Essentially the body takes part in figurations. We are aware of this distinction when we for instance say: 'I am born in . . . , but I grew up in . . .' where we also signal a priority and indicate the condition that we find most important. Hereby, we indicate a state of bodily tension and vivacity, meaning readiness and attitude towards one's surroundings which, as the American body phenomenologist Alphonso Lingis expresses it, encircles far more aspects of man's embodiment than the aspect of mere presence (Lingis 1991: 6,9,13). The body does not tend to a state of rest; rather it is the body's ability to ex-istence, i.e. the transgression of mere standing, state and stance, which, in moving, enables the body to display potentiality and openness towards its surroundings. The ex-static and open body is symbiotic, relational and unfinished. The ex-stasis refers to both the outstanding body and the moveable body, whose openness and relationality seems evident when regarding the fact that although not able to look round the corner, you can certainly both smell and listen round it. This includes the transgression of mere being, pointing instead at the body's ability to both be-longing and longing by means of the senses mingling with the world by means of reaching out.

Moreover, this body as a means of self-knowledge, does not occupy space, but rather inhabits space, and this space is not measured by a spatiality of position, but is characterised as lived space or as spatiality of situation (Schrag 1997: 55). Consequently, the identity of such a body is more a question of 'who' than 'what', because identity is elaborated through actions, involvements and experiences much more than being the outer expression of changeless essence and rigid immutability. This distinction between the state of being and belonging appears even more clearly when we hold another dimension of place, place as history place.

This third place is personal, closely related to one's life history and memories. However, these memories first of all concern activities, relations and possibilities. The historical identity place is first of all a virtual place, and it is historic, not by laborious history-making, but simply by living the history here by virtue of actions and doings, where leisure gets still more relevant, because it implies a wider range of bodily life-historical actions and experiences than work. Leisure implies the possibility of acquiring knowledge of places in an everyday geography.

But the link between place and history is as much about history in a wider sense as it is about national and homeland aspects, where you transcend the everyday level of active geographical knowledge. In this connection experience as well as mythical and fictitious factors are part of the historical dealing with a place. Bodily-physical dimensions are part of this. The historical belonging to a place in either biographical or national regard is not only of a mental nature, as we will explore in particular examples of leisure practice. Even the 'official' historical space, claiming respect and deference, is also a space to move in, to socialise and have experiences in. It is not until you move and perceive in such historical places that they become a part of your own history. Only then do they become a part of your overall or national historical home, while on the other hand, they are not real historical places until perceived and appropriated by living and moving people. Wembley would never have been a national historical place and a mecca of English soccer if it had always stood empty. Both the national soccer team and the spectators have created it. Correspondingly, the Lake District would have remained a geological element had it not been for the farmers and tourists. Leisure facilitates both bodily autonomy and sensory heteronomy.

Place and home

This differentiation between three different dimensions of place can be considered from the phenomenon of childhood home, 'birthplace', i.e. the concrete version in the form of household, family, house, face-to-face relations (Hobsbawm 1993: 61–4, Rykwerth 1993: 47–58). The identity connected with birthplace is especially strong in childhood but when adults mention their home and tell us where they come from, they refer not to a particular house, but to a place in a wider sense (Hobsbawm op. cit.).

In the Danish language this 'home' is characterised as 'a community', a part of the country that is both our own and at the same time so comprehensive that it contains many possibilities of relations and experiences and goes beyond individuality to point at collectivism, or at the connection between individuality and collectivism. The community is just so strong that it lets the individual into the community without forgetting the characteristics of the individual. Hobsbawm characterises this shift as a change from 'Heim' to 'Heimat', pointing at the sociability of Heimat, in which the potentiality of relations and the range of experiences (mental and sensory) is enhanced when compared with the enclosed home of the hearth. In his book on the football grounds of Great Britain, Simon Inglis considers a picture of an aged faithful Charlton fan picking up a part of the pitch of 'The Valley' as a souvenir. The important thing is not the particular piece of turf but rather his picking it up. You would never say that Charlton Athletics' ground is his home, but it is most certainly a part of his Heimat, not least due to this appropriating act, evident in the presence of people in the background smiling at this touching and moving Heimat-making.

The Heimat relation has only limited reach. If we are far away, abroad for instance, we are rather dealing with a third and more comprehensive level, namely one's country, one's nation, i.e. the third level, the national historical level. But even here it is obvious that there is a bodily action-dimension included. There is much affinity between the cricket landscape and the emblematic presentation of English eternalism. The cricket ground, not least situated in a rural southern landscape with a Kentish village stereotype, symbolises the real England reflecting English values by means of not only the players, but also the spectators sitting in the outer parts of the grounds in deck-chairs or on wooden benches, breathing the rural air with the smell of haystacks and hedgerows (Bale 1988). The *leitmotif*, clearly, is security and contentment, where an indelicate factory chimney hardly dares to spoil the rhododendron air, but it is also activity and bodily pleasure in partaking in a landscape together with other people.

The concept of home seems, in other words, to be as closely connected with relational leisure activities, as it is relative to both time and space; the meaning of it seems to be contextual (Hobsbawm 1993: 63), as can also be seen in the revitalisation of the gaúcho culture of Rio Grande do Sul in southern Brazil. Here it is striking that this revitalisation as a distinct leisure activity is sustained by movement and perception such as music festivals, maté drinking, horse riding, dancing (Oliven 1996). Such activities are not mirrors of an identity reflecting the region of Rio Grande do Sul as home; rather they are leisure activities creating a relational context to be part of, by means of common bodily actions and experiences, where you are both an active or originating force and a reactive and responding force.

281

This implies that identity is constructed by partaking in 'identity work' in leisure time, Here identity is created by means of a practice that does not necessarily have 'identity' and 'home' as an aim, but rather as a gratuitous concomitant acquired silently as tacit knowledge. This happens through routines, habits and invented traditions, that are all characterised by action and practice, where you acquire knowledge by being rapt in the spatiality of the situation.

Having established the three space dimensions, or rather place dimensions, and three forms of home place, all embedded and sustained through leisure practices, we can now continue. We shall do that by examining how the space that we, in our leisure, relate to with such confidentiality that we call it our home place is built up, hoping to be able to isolate a common denominator repeated at all three levels as discussed above from different angles.

The anthropological research in 'sense of place' represented by Marc Augé is informative (1995). He points to the fact that the home space primarily consists of a geometric pattern as there are roads, axes and paths that lead from one place to another and link people round the established and official canals. Such geographical patterns are dug out unofficially by means of our feet, vehicles. There are also official crossroads, street corners and open squares where people meet, shop, chatter, tell stories, fight at night etc. Finally, in the life history place there are centres of a more or less monumental kind, often connected with either a religious or a political representation in relation to which other people feel different. But also here the history is primarily attached to leisure movement, i.e. movement of a searching, touristic or travelling nature. In such cases the life historical as well as the official historical aspects of the place converge.

In each case of 'home' there is movement and activity. In the movement and the potentials of our embodiment we find the common denominator repeated in the place relation that makes us define a place as our own. An analysis of the history of the space must therefore implicate the history of movement and sight as it is movement that creates space (Löfgren 1992: 189).

Topophilia and homotopia

The modern tendency for state and commodification to obliterate differences and create similarities, especially as regards sense of place and locality, leads to an increased unification and levelling of an increasing number of localities – and by that a lack of possibilities for the sense of home.

Marc Augé mentions this unification as a tendency to create 'non-places', or a Foucauldian 'homotopia' (Bale 1996: 233–50). An expression of homotopia is for instance seen in the landscape of the bypasses and wholesale stores, diners, streets, parking lots, motorways, supermarkets, marinas and in the glamorous no-man's land of the airports (Augé 1995: 1). They appear as anonymous and dedifferentiated 'non-places' which look the same all over the world. Marc Augé has pointed out that they are characterised by being without qualities of place, thus without relations and without history (1995: 77). Juxtaposed to this de-differentiation we can, with the idea of *la poetique* (Bachelard 1957), emphasise the concept of topophilia as a name for love for and affiliation to a certain place (Tuan 1977, Bale 1993: 64–77), adding to Augé's analysis of place.

Topophilia cannot be of an abstract character, but is always connected to a specific place; it is localised; this is the first and most important quality. At the same time it is the sense of place that differentiates it from a similar longing after rituals. Admittedly, rituals are characterised by the combination of space and movement, but the activity of ritual need not attach to a place in the same manner; it can happen at any place because the important thing here is move-ment and activity, not the combination of place and movement. Topophilia is attached to the dimension of space which has to do with place when there is a felt relation.

This does not necessarily need to be attached to a specific locality, rather a *type* of place, for example at the intersection of a unique simple locality and the common space, but it still has its own characteristics. It is important, however, that it invites bodily activity and perception and gives the impression that it receives you so that you feel it to be a home.

Place and horizon

Places with a distinct horizon can confer a poetics of home, poised between what is certain and what is possible. The landscape of the horizon always points to the possibilities and is characterised by being a 'not-yet-place' (Kayser Nielsen 1994: 64–5). This is the case whether you walk along the coast of Western Jutland with a ship high up on the horizon or on the slopes of the Pennines with sheep high up near the crest. At the same time as the horizon encapsulates the lim-itation of space and lets us sense the possibilities behind the horizon, the horizon also points to the possibilities of change in space. If we move a few steps further, the perception of space is changed immediately. We know this to be so, it can be anticipated. Thanks to the horizon we have the possibility of experiencing ourselves phenomenally as bodies in a space, a space that due to the fact that we are moving is attached to the possibilities of change. What we are experi-encing is the relativity of the space (Kern 1983: 131–181). So we are attached to the control of space, but we also contribute to the sensory version of this control of space.

This is probably the reason why it is so attractive to stay by the sea with a very clear relation between horizon and fixed place (Kayser Nielsen 1993: 7–23). Here for example, at leisure at the seaside, we have the opportunity to feel

Plate 11 Do be do – be do – or not to do?

Photo: Niels Kayser Nielsen

ex-static, because the horizon, as a bodily experienced vision, points at new possibilities and challenges. Once again it is obvious that the body as a dimension of selfhood does not occupy space, but inhabits space. At the horizon, and at the horizon by the sea in particular, we experience ourselves as a part of 'nature' but as soon as we move, we realise that we are capable of knowing that space in a new way. We are able to create a new space by moving.

A horizon landscape is not only attractive on account of the phenomenal sensory observation of the fundamental conditions of the human body, but also because of the social understanding of this condition. The horizon is the symbolic expression of our ability to make plans and projects (in Latin, *projectere* means 'to throw out') in contrast to the actual situation. As the Danish painter Oluf Høst once wrote in a diary aphorism: 'Behind the horizon – but not forgotten – live your life without letting your dreams go' (Høst 1967: 50), or, with a quotation from one of the German rock-poet Udo Lindenberg's songs: 'Hinter'm Horizont geht's weiter' ('It will continue behind the horizon'). The other side of the crest does not necessarily mean 'over the hill'; it can also signify the place

284

of 'otherness' and utopian possibilities, while at the same time the horizon line points back at you, situated in a certain place, in a certain limited viewpoint.

The horizon 'landscape' is, in other words, the place where you might see yourself as belonging and where, on the other hand, the pot of longing is kept boiling. In this way it is also a symbolic place where our limitations and potentials are in focus. But while the horizon places seem fascinating due to their demonstration of very common, bodily conditions, there are other places that seem to a much higher extent to be connected to our specific social and cultural conditions.

National homes – at least in Norway

The Norwegians are, according to both their own and other people's understanding, very pleased about 'nature'. But why? Being happy is not an abstract idea in Norway, (Enzensberger 1988: 207). Happiness in Norway is made of wood, grass, rocks and salt water. A Norwegian survey made in the 1980s asked: If you had more leisure time what three activities would you spend more time on? Norwegians prioritised walks in the woods and in the country; skiing and hiking; exercise and sports. Norwegian anthropologist Tove Nedrelid started wondering why this focus on 'nature' was so (1991).

One possible explanation could be the fact that the Norwegians seek the open air because they are used to it, because habitually it has become their 'second nature' to do so. Another explanation could be the traditional utilisation of nature in the form of gathering berries. It sounds reasonable in itself, but can hardly be sufficient as Norwegians are not dependent on a self-sufficiency economy. It is not about a materialistic use of nature. Nedrelid is therefore of the opinion that the Norwegians wish to get out into the open air because of the bodily and mental joy of outdoor activities, including berrying. It is less a contemplative and aesthetically oriented passive enjoyment of 'nature' but rather a bodily way of combining it with leisure as a means of both creating and confirming identity as a home place experience.

The picking of berries has a double character: it is both serious and playful, it is both a question of having a good grounding and some variety (Edvardsen 1997: 105). It seems that grass, rocks, trees, water – and the experiences of the mouth and the kitchen as focus of our world – are the kinds of constituent that are important in the make-up of leisure. And it also seems that they are important preconditions for gaining knowledge about yourself as being both situated and ex-istent in a home place. The physical landscape triggers knowledge and leisure is an important way to release these components of contentment. The landscape offers leisure activities that provide an opportunity to gain knowledge, not only of the landscape as external object, but also of the self as being in the world, as belonging to a certain, local place, where there is correspondence between your bodily finality and a certain specific landscape. This correspondence would be impossible in an abstract 'non place' where no relationship with the self is felt.

Physical activity out in the open air seems to be a way to acquire 'Norway'. In Denmark this may take the form of a national place like 'Himmelbjerget' (the Sky Mountain) in Jutland *by climbing it and not only thinking about it*. Having a nationality like that, at least in Scandinavia, often means physical presence in a landscape. To move physically in a landscape, to get out into the open air, to use the body, becomes a work of identity that we undertake with the national popular purpose of self-understanding (Olwig 1994: 154–69). Spending the vacation caravanning in Denmark is not only a question of not being able to afford a charter vacation to the south; it is also a question of 'getting to know your country', very often together with other people with whom you share a common space, thereby making it a place. It is understood that we do not get to know our country by thinking of it or by sitting in the armchair, but by collective physical contact. The Finns who go out to their cottages at the sea or at a lake do not necessarily spend their leisure time with an explicit Finnishness in mind, but through their leisure activities, such as sauna bathing, hunting or fishing they express Finnishness, not as an idea but as a competence acquired through activity and outdoor life (Kayser Nielsen 1997a: 125–26).

As Orvar Löfgren has said, it is a misunderstanding that space and landscapes are something that we enter and then leave again. Rather, as we saw it with regard to the gathering of berries in Norway and the sauna bathing in Finland, that landscape and space is something we create, activate and articulate with the activities we perform in the space. Landscape and space is created through use (Löfgren 1992: 189). So, the ruling paradigm for the topography of home in Scandinavia is movement, partly in contrast to being pastoral in Britain and zoological/eugenic in Germany (Schama 1993: 7).

Home as criticism of 'purity'

The character of the home is contingent, cannot be generalised. It is a mark of what Clifford Geertz calls 'local knowledge'; a contextualised and specific knowledge that is attached to local conditions and is dependent on the circumstances. It finds expression in so called traditional societies' medical diagnoses and in questions of legal and moral responsibility, discriminating between universal 'law' and local 'rights': sailing, gardening, politics and poetry, law and ethnography are 'crafts of place' (1993: 167). The idea of home is always attached to a relation between the participants involved and the unique circumstances in question. Statements about home only have validity in relation to an audience with definite preferences and cannot, from a substantial point of view, have a universal and general validity. Home and nation get their power from their spatiality, but to be more precise: the ideas of home and nation emerge from specific and lived space, not from imagined space (Löfgren 1997: 33). Leisure activities provide a means of bodily access to live in space, or rather to make space living by ex-istence at a certain place that you appropriate through activity and sensory knowledge. As the Danish author Johannes V. Jensen wrote in the novel *Kongens Fald* (The Fall of the King), describing a lived- and a living-space; i.e. a place:

> After all it was his realm. He saw in a vision Denmark, a reality in the sea, a sum of areas in all colours, a *land*. It is forever true that Denmark lies between the two blue seas, green in the summer, rusty in the autumn and white under the wintry sky . . . The sun stands in fans over the hills at Limfjorden where the west wind blows homelike.
>
> (Jensen 1968: 136)

This concept of home is related to both very early and very late modernity. Leisure practice would seem to exemplify late modernity in its knowledge complexity and contextual dependency, body, movement, and sensuality in everyday life. It can definitely be read as an attack on scientific knowledge, as a defence of kaleidoscopic multiplicity and contextual dependence in the 'purity' with a sense of normality, conformity and a panoptical discourse, and existence of human beings (Toulmin 1990). It means to him a defence of the oral pecu-liarity in preference to the written, of the particular in preference to the ordinary and of phenomena attached to time in preference to what is without time. Abstract generalisations are to him a nuisance and his main opponent therefore becomes Descartes with his search for abstract and general ideas that must not be contextual and dependent on such contingent factors as body, movement, sensuality – and such tedious factors as place, weather and landscape.

Abstract axioms should, according to Descartes, not be determined by this kind of coincidence but ought to be thinkable no matter whether we find ourselves in Königsberg, in Oslo or in Sicily. Abstract axioms are independent precisely of time, space and place. As a consequence they do not have a place of home but distinguish themselves on the contrary by being of universal validity. It also means that they, according to Yi-Fu Tuan, are 'dead in a way as they can only exist at the expense of sensuous commitment to particular individuals and local-ities' (1996: 947).

Conclusion

In conclusion, we must hold on to the fact that the feeling of home is inex-tricably related to a basis of experience, including specification and bodily presence, a basis which leisure, in contemporary society, is apt to provide. Home place will never be able to develop into an axiom in a traditional Western philosophical sense, in the meaning of universal validity – unless, of course, it develops into a crypto-religious dogma. But this is exactly what the idea of home place is not suitable for, because specification, locality and dependence on time are attached to it.

The feeling of home is based upon movements, experience and physical activities. It is therefore individual and diffuse, 'unclean' and contingent, but therefore also more durable than the firm nationalism with its demands for borders, system and order. The contingence and the missing 'Ordnung' of home is too blurred to be able to bring people together in common, ideological passion. This interpretation of home emphasises how important it is to cultural studies (to which geography certainly belongs) to take both the possibilities of leisure and the condition of embodiment as a starting point. This starting point will facilitate humility and lack of will to grandness and firmness. But it will also facilitate our understanding of how knowledge has bodily experiences as grounding.

References

Augé, Marc (1995) *Non-Places. Introduction to an Anthropology of Supermodernity*, Verso: London.

Bachelard, Gaston (1957) *La Poétique de l'espace*, Presses Universitaires de France: Paris.

Bale, John (1988) 'Rustic and Rational Landscapes of Cricket', in *Sport Place International. An International Magazine of Sports Geography* 2(2), Fall.

Bale, John (1993) *Sport, Space and the City*, Leicester University Press: Leicester.

Bale, John (1996) '"Homotopia". The Sameness of Sports Places', in Henning Eichberg and Jørn Hansen (eds), *Bewegungsräume. Körperanthropologische Beiträge*, Afra Verlag: Butzbach-Griedel.

Bausinger, Hermann (1980) 'Heimat und Identität', in Elisabeth Moosmann (ed.), *Heimat. Sehnsucht nach Identität*, Ästhetik und Kommunikation Verlag: Berlin.

Berman, Morris (1990) *Coming to our Senses. Body and History in the Hidden History of the West*, Unwin: London.

Edvardsen, Edmund (1997) *Nordlendingen*, Pax: Oslo.

Enzensberger, Hans Magnus (1988) *Åh Europa*, København: Gyldendal.

Geertz, Clifford (1993) *Local Knowledge. Further Essays in Interpretive Anthropology*, Fontana Press: London. Originally published 1983.

Hobsbawm, Eric (1993) 'Introduction', in Arien Mack (ed.), *Home. A Place in the World*, New York University Press: New York.

Høst, Oluf (1967) *Fra Orion til Gudhjem*, Hans Reitzel: Copenhagen.

Inglis, Simon (1993) *The Football Grounds of Great Britain*, Collins Willow: London.

Jensen, Johannes V. (1968) *Kongens Fald*, Copenhagen: Gyldendal. Originally published 1900.

Kayser Nielsen, Niels (1993) 'Fare, frihed og sundhed – om vandforestillinger i moderne tid', *Den jyske Historiker* 65.

Kayser Nielsen, Niels (1994) 'Sport, Landscape and Horizon', in John Bale (ed.), *Community, Landscape and Identity: Horizons in a Geography of Sport*, Keele University: Department of Geography. Occasional Paper no. 20.

Kayser Nielsen, Niels (1997a) 'Att tillägna sig det nationella i Norden genom friluftsliv och hälsa', in Maria Fremer, Pirkko Lilius and Mirja Saari (eds), *Norden i Europa. Brott eller kontinuitet?* Institut för nordiska språk och litteratur: Helsingfors.

Kayser Nielsen, Niels (1997b) 'Movement, Landscape and Sport. Comparative Aspects of Nordic Nationalism between the Wars', *Ethnologia Scandinavica*: 84–98.

Kern, Stephen (1983) *The Culture of Time and Space 1880–1918*, Harvard University Press: Cambridge, Mass.

Lingis, Alphonso (1991) *Foreign Bodies*, Routledge: New York and London.

Löfgren, Orvar (1992) 'Rum og bevægelse', in Kirsten Hastrup (ed.), *Den nordiske verden*, vol. 1, Gyldendal: Copenhagen.

Löfgren, Orvar (1997) 'Att ta plats: rummets och rörelsens pedagogik', in Gunnar Alsmark (ed.), *Skjorta eller själ? Kulturella identiteter i tid och rum*, Studentlitteratur: Lund.

Morley, David and Kevin Robbins (1993) 'No Place like Heimat', in Erica Carter, James Donald and Judith Squires (eds), *Space and Place. Theories of Identity and Location*, Lawrence and Wishart: London.

Nedrelid, Tove (1991) 'Use of Nature as a Norwegian Characteristic. Myth and Reality', *Ethnologia Scandinavica*: 19–33.

Oliven, Ruben (1996) *Traditional Matters: Modern Gaúcho Identity in Brazil*, Columbia University Press: New York.

Olwig, Kenneth R. (1994) 'Landscape, landskab, and the body', in Anders Linde-Laursen and Jan Olof Nilsson (eds), *Nordic Landscapes. Cultural Studies of Place*, Nordisk Ministerråd: Copenhagen and Stockholm.

Rykwerth, Joseph (1993) 'House and Home', in Arien Mack (ed.), *Home. A Place in the World*. New York University Press: New York

Schama, Simon (1993): 'Homelands', in Arien Mack (ed.), *Home. A Place in the World*, New York University Press: New York.

Schrag, Calvin O. (1997) *The Self after Postmodernity*, Yale University Press: New Haven and London.

Toulmin, Stephen (1990) *Cosmopolis. The Hidden Agenda of Modernity*, University of Chicago Press: Chicago.

Tuan, Yi-Fu (1977) *Space and Place: The Perspectives of Experience*, Minnesota University Press: Minnesota.

Tuan, Yi-Fu (1996) 'Home and World, Cosmopolitanism and Ethnicity. Key Concepts in Contemporary Human Geography', in *The Companion Encyclopedia of Geography*, Routledge: London.

NAME INDEX

SUBJECT INDEX